Bird Census Techniques

Second Edition

Bird Census Techniques

Second Edition

Colin J. Bibby
BirdLife International

Neil D. Burgess
Danish Centre for Tropical Biodiversity

David A. Hill
Ecoscope Applied Ecologists

Simon Mustoe
Ecoscope Applied Ecologists

Illustrated by Sandra Lambton, RSPB and Simon Mustoe

ACADEMIC PRESS

A Harcourt Science and Technology Company

London • San Diego • New York
Boston • Sydney • Tokyo • Toronto

Copyright © 2000 by ACADEMIC PRESS

First edition published in 1992

Academic Press
A Harcourt Science and Technology Company
Harcourt Place, 32 Jamestown Road, London NW1 7BY, UK
http://www.academicpress.com

Academic Press
A Harcourt Science and Technology Company
525 B Street, Suite 1900, San Diego, California 92101-4495, USA
http://www.academicpress.com

ISBN 0-12-095831-7 Hardback

Library of Congress Catalog Card Number: 00–101511

A catalogue for this book is available from the British Library

Designed and typeset by Kenneth Burnley, Wirral, Cheshire

Printed and bound by CPI Antony Rowe, Eastbourne

Acknowledgements

For this second edition of the book, we would like to thank a number of institutions and people who have made a notable contribution to its completion. Firstly we would like to thank our institutions who have allowed us to spend some time on the task and have allowed the use of their facilities. These institutions are Secretariat of BirdLife International based in Cambridge, UK (Colin Bibby), the Danish Centre for Tropical Biodiversity located in the Zoological Museum of the University of Copenhagen, and in particular Prof. Jon Fjeldså (Neil Burgess), and Ecoscope Applied Ecologists (Simon Mustoe and David Hill).

The following people were particularly helpful in the updating of the various chapters: Knud Falk and Henrik Skov (Ornis Consult, Copenhagen), Jon Fjeldså, Louis Hansen, Lars Dinesen, Poul Andersen, Bent Otto Poulsen, Kaj Kamp and Carsten Rahbek (Zoological Museum in Copenhagen), Len Thomas and Steve Buckland (Research Unit for Wildlife and Population Assessment, University of St Andrews), Rhys Green (RSPB/Cambridge University), Derek Pomeroy (Makerere University, Uganda).

New permissions were also obtained for figures in the book and we would like to thank the following for allowing the reproduction of these: Neumann Verlag Radebeul, Royal Society for the Protection of Birds (UK), Joint Nature Conservation Committee of the UK Government, Seabird Group (UK), Wetlands International (Kuala Lumpar Office), Finnish Game and Fisheries Research Institute and Scottish Natural Heritage. We would particularly like to thank the University of St Andrews Research Unit for Wildlife and Population Assessment, for allowing us to reproduce screenshots of the new Distance interface.

We would also like to thank the following people who made such a large contribution to the first edition of the book, upon which this revision is based: Sandra Lambton, Adrian del Nevo, James Cadbury, Leslie Street, Dave Allen, Geoff Welch, John Wilson, Diana Ward, Lynn Giddings, Ian Dawson, Anita McClune, Gareth Thomas, Graham Hirons, Ceri Evans, John Cayford (RSPB, UK), Jeremy Greenwood, Rob Fuller, John Marchant, Chris Mead, Will Peach, Steve Carter, Paul Green, Rowena Langtson (BTO, UK), Jeff Kirby & Peter Robertson (Wildfowl and Wetlands Trust, UK), Dick Potts and Steve Tapper (Game Conservancy Trust, UK), Robert Petley-Jones (English Nature).

The following publishers also gave permission for the original figures, many of which are used here as well: Academic Press Ltd, American Ornithologists Union, British Trust for Ornithology, Blackwell Scientific Publications Ltd, British Birds Ltd, British Ecological Society, Cambridge University Press, Canadian Wildlife Service, Chapman and Hall, Cooper Ornithological Society, Devon Birdwatching and Preservation Society, Game Conservancy Trust, Gauthier-Villars, Harcourt Brace Jovanovich, Institute of Terrestrial Ecology, Macmillan Magazines Ltd, Ornis Scandinavica, T & AD Poyser Ltd, Royal Society for the Protection of Birds, The Wildlife Society, Wildfowl and Wetlands Trust, US Fish and Wildlife Service.

About the Authors

Colin Bibby is Director of Science and Policy at BirdLife International where he is interested in the role of birds as indicators for global biodiversity conservation. While Head of Conservation Science at the Royal Society for the Protection of Birds, he was a co-author of *Red Data Birds in Britain*. In both capacities, he has been struck by the small number of birds of conservation concern which have been counted adequately. He has counted birds in Britain and Europe, as a professional, as an amateur participant, and as an organiser of surveys for the British Trust for Ornithology. He was motivated to start this book by the belief that bird-watchers would contribute more to conservation if they put more effort into counting birds, but lack of guidance on methods was a handicap.

Contact: BirdLife International, Wellbrook Court, Girton Road, Cambridge CB3 0NA, UK.

Neil Burgess completed a BSc in Botany and a PhD in Palaeobotany (evolution of the earliest land plants), after which he started work at the Royal Society for the Protection of Birds. He worked on reserves and in the headquarters in Sandy on the management of habitats for birds. During this time the main work on the first edition of *Bird Census Techniques* was completed. In 1990 he became more involved with African conservation and biodiversity research through project management in the Africa Section of the International Department of the RSPB, and as Research Director of the Society for Environmental Exploration. Most of his work concentrated on tropical forest and wetland habitats in Africa and SE Asia, including ornithological surveys. He moved to the Danish Centre for Tropical Biodiversity, at the University of Copenhagen in 1994, to coordinate and develop programmes aiming to map the distribution of biodiversity in Africa. He also worked as International Project Development Officer for the Danish Ornithological Society in 1995–1999. He has maintained an interest in ornithological survey work in Europe, but has moved increasingly into project management work in the tropics and eastern Europe mainly through setting up and managing programmes to assess biodiversity values and to undertake practical conservation actions.

Contact: Danish Centre for Tropical Biodiversity, Zoological Museum, Universitetsparken 15, Copenhagen Ø, DK-2100, Denmark.

David Hill is Chief Executive of Ecoscope Applied Ecologists which was established in 1992 to provide research and consulting services in ecology to both the public and private sectors in the UK and overseas. Much of Ecoscope's work involves the use of quality survey methods, many based on those described in this book. These methods are becoming increasingly important as standards for environmental assessment and he is actively involved in developing and setting practice standards for survey and evaluation within the profession. David Hill received his doctorate in bird ecology from the Edward Grey Institute of Field Ornithology at the University of Oxford in 1982, after which he moved to the Game Conservancy and later, as Senior Ecologist, to the Royal Society for the Protection of Birds. Much of this work involved studies of marked birds, their behaviour and habitat preferences in relation to land use. In addition, he set up experiments and monitoring studies on RSPB reserves which brought him into close contact with sampling different bird species and habitats. In 1989 he became Director of Development at the British Trust for Ornithology with responsibility for setting up and running research contracts on estuaries, farmland, woodland and uplands, using the methods outlined in this book. He has published a number of books and numerous papers on bird population ecology, habitat ecology and conservation, ecological assessment and mitigation. What started as a passion at the age of five, has provided a rewarding career. We hope this book, in giving others tools to count birds, will engage enthusiasm and provide a similar reward to many, whilst at the same time yielding essential information for conservation.

Contact: Ecoscope Applied Ecologists, Crake Holme, Muker, Richmond, DL11 6QH, UK.

Simon Mustoe is a research consultant with Ecoscope Applied Ecologists. He graduated in Ecology from the University of East Anglia in 1995, after which he led a multi-national team of researchers to Madagascar as part of an integrated conservation development project, part-funded by the BirdLife/Fauna and Flora International/British Petroleum Conservation Programme. He has worked both as a bird researcher at the Royal Society for the Protection of Birds and as an assistant to the African Important Bird Areas (IBA) project at BirdLife International. Simon specialises in ornithological assessments, being familiar with a wide range of different methods and having worked in a range of situations as diverse as tropical forests and the north-east Atlantic. His experience of modern methods of large-scale surveys using distance sampling led to his being involved in the important task of bringing these chapters up to date.

Contact: Ecoscope Applied Ecologists, 9 Bennell Court, Comberton, Cambridge CB3 7DS, UK.

Contents

Preface to the Second Edition

Bird Census Techniques has been pivotal in raising the profile and standard of bird surveying worldwide, which has contributed to its ongoing success and popularity in many parts of the world. Since the first edition, however, existing bird monitoring methods have been refined and new methods have been invented. Exciting new software has been developed to simplify the relatively complex modelling and analysis of data as diverse as mark–recapture and line transect sampling. New maps showing the distribution of birds on national and continental scales have been produced and monitoring schemes have been established allowing us to assess the associated trends in bird numbers and density. Worldwide, bird monitoring has revealed a shocking decline in the numbers of many formerly common bird species. The reasons for these population declines are intensely debated, but the intensified use of land by mankind is a primary factor for many species. This has given rise to several independent lists of bird status on National, Continental and Global scales, which are now important factors in deciding the outcome of land-use issues, especially those relating to development. The methods described in this book are now being applied, for example, to Ecological Assessments and Environmental Impact Assessments, thereby increasing their detail and scientific quality.

The first edition of this book was conceived as a tool to amalgamate text on the various bird counting methodologies and act as a handbook for ornithological research. This aim has not changed, although the extent of the literature has increased dramatically. This second edition is offered to professional and amateur researchers, volunteer conservationists, consultant ecologists, and anyone else who is planning to survey and monitor birds but who may lack the facilities to research and understand the bewildering range of modern survey methods. We hope the guide will also benefit students who are studying the varying and often conflicting bird census literature and that this edition might help the reader understand the problems of bird counting and thereby devise competent and valuable experiments in a clear and cost-effective way.

1

Purpose and Design in Counting Birds

Introduction

This chapter is devoted to the importance of bird censusing and how it may be applied to conservation. It also discusses the types of questions you will need to ask before embarking on a project, which methods are best suited to meet the objectives of a survey and which are appropriate to the type of species or environment being studied?

Each case refers to chapters from the rest of the book which can be further consulted where necessary. Chapter 2 considers aspects of survey design that minimise bias and increase the accuracy of results. We only introduce the concept of statistical theory; for more detail we suggest referring to Zar (1984) and Fowler and Cohen (1986, 1990). Chapters 3–7 further explore the theory and application of methods to survey birds, including territory mapping, line transects, point counts, species richness and diversity, and capture and marking. In each chapter, we have investigated and reviewed data interpretation and analysis with reference to some of the best (and worst) examples. Chapters 8 and 9 give an insight into methods of surveying various species and groups of species and how methods have been adapted to meet the requirements of the survey. Chapter 10 documents the theory and application of distribution studies and Chapter 11 explores methods of assessing bird habitat.

Why is bird censusing important?

Birds are relatively easy to census as they are well known, easily recognisable and simpler to locate than many other taxonomic groups. Birds can be useful indicators of the state of the environment and are also key species for education and public awareness. For example, publicity surrounding investigations into the population decline of farmland birds in the UK, has turned the Skylark into a national conservation icon.

There is now a considerable knowledge of bird numbers and distribution in some countries. A complete list of avian population estimates was recently published for Britain and the United Kingdom (Stone *et al.* 1997) and the *New Atlas of Breeding Birds in Britain and Ireland* (Gibbons *et al.* 1993) identified population and distribution changes since the first atlas was published 17 years before (Sharrock 1976). Reliable baseline data makes bird monitoring even more important as there is a real chance of identifying future changes in distribution, abundance and diversity.

Once you have a dependable baseline estimate of bird distribution and abundance, monitoring can identify changes in population and range in relation to environmental features such as habitat and land-use, and direct threats, e.g. persecution. Bird census techniques are now used extensively in Environmental Impact Assessments (EIAs), and to identify success or failure of conservation management schemes in protected areas.

In respect of EIA's, the new Environment Assessment Regulations place greater emphasis on planning authorities to demand good environment information as part of a planning application for a development. Bird surveying is therefore directly linked to legislation surrounding development of the environment. As part of an EIA, the area proposed for development may be surveyed to identify important populations of birds and where they are distributed. It is then possible to decide whether the development is likely to have an impact on species of particular importance. In addition, some species may be specially protected by law, in which case, development may be halted, at least until sufficient mitigation measures are in place. Correct and adequate bird censusing therefore has significant financial leverage and importance.

The ecological part of an environmental impact assessment, where birds have been determined as a potentially important component, should aim to ensure that new areas are surveyed for rare species and that where possible, species diversity is maintained or enhanced by development. If the site is identified as particularly important, then it may be a candidate for designation as a protected area. In such cases, planning permission is usually denied, at least until a more thorough study is done and a final decision is made. In the United Kingdom, developers are obliged to meet any requirements for environmental management of the site by signing up to a Section 106 Agreement of the Town and Country Planning Act, or other legally binding instrument. Failure to achieve these predetermined standards should ideally be identified by post-development monitoring of the site. Similar laws exist in many other countries.

The criteria by which species are designated 'important' or not may be described in national and international lists of species of conservation importance. For example, in the UK, there are two main lists: the Joint Nature Conservancy Council's (JNCCs) *List of Birds of Conservation Importance* (JNCC 1996) and the Non-Governmental Organisations *Red List of Birds of Conservation Concern* (Gibbons *et al.* 1996). Both lists use criteria based on national population decline and range contraction, European and Global status. Global status is listed in BirdLife International's *Birds to Watch 2* (Collar *et al.* 1994), which categorises world bird species in IUCN threat categories, based on evidence of range, population size and level of threat. European status is established in a separate list of Species of European Conservation Concern (SPEC), detailed in *Birds in Europe* (Tucker and Heath 1995).

Once you have identified that a site contains an important number of species, the area may be a candidate for special protection. For many years, the UK has had a network of Sites of Special Scientific Interest (SSSIs). In different habitats, birds are given importance 'scores'. For example, Black-necked Grebe in lowland open waters scores 5 points. If the total number of points achieved on a site exceeds a threshold level, then the site can be considered for SSSI designation.

The European Union Wild Birds Directive (79/409/EEC) dictates that areas regularly supporting at least 1% of the global or biogeographic population of a species or subspecies of waterfowl, or areas that regularly support concentrations of greater than 20 000 birds, are internationally important. Sites that support 1% of the national population of a species are deemed nationally important. Such sites would qualify as candidate Special Protection Areas (SPAs). In the case of SPAs, the government is legally bound (through European legislation) to ensure that certain conservation criteria are met and that the level of conservation importance does not deteriorate.

Choosing a method

Choosing an appropriate method is easier if the study has clear, predetermined aims. It is a mistake to rush to the field and start counting without first deciding which method will provide the most suitable results. A project may not meet its full potential unless these aims are properly understood and researched before data collection begins.

A few recent publications (see below) explore generic methods of bird survey, and more particular ways of assessing the abundance and distribution of species or groups of species. For anyone mounting a survey, it may be useful to refer to such texts and contact national bird organisations that are already involved in monitoring. It is also important to ensure that the species is not legally protected. In Britain, special protection under Schedule 1 of the Wildlife and Countryside Act makes survey of species such as breeding Barn Owl illegal at or near the nest site, except under licence from the government's statutory conservation bodies. In any case, it is advisable to research the behaviour of the bird and exercise care to minimise disturbance, particularly if the survey involves approaching nests.

Before planning your methods consider the variety of possible scenarios that could dictate your project's fieldwork techniques. Ask whether your results are intended to apply to a wide geographical area or to a single site? Are many species involved or just one? Are accurate counts needed or will relative counts or presence and absence data suffice?

The available man power for counting is usually limited and may only be adequate for studies of a single site or species. Absolute (exact number) counts are often very difficult to obtain but for many purposes are not really needed. The key to a good study lies in recognising what kinds of data are required and any weaknesses inherent in the methods. It is easy to agonise over achieving statistically precise results when it may often be enough that estimates are right to within 100–200%. Considerations about the accuracy of methods and the precision of results are so important that Chapter 2 is devoted to them. In truth, the perfect method for counting birds does not exist, but this need not prevent the extraction of useful results from a well prepared study. For a review of data analysis and the design of experiments in ornithology the reader is referred to James and McCulloch (1985) and Hairston (1989).

The following publications are regularly referred to in this book. For a comprehensive review of the statistics and theory behind distance sampling methods, we suggest reading *Distance Sampling: Estimating the Abundance of Biological Populations* (Buckland *et al.* 1993). For a guide to standard monitoring of species or existing methods of monitoring groups of species in the UK, refer to the RSPB's new *Bird Monitoring Methods: a manual of techniques for key UK species* (Gilbert *et al.* 1998). Surveying seabirds in colonies or at sea is covered by Walsh *et al.* (1995) and surveying by ship or plane by Webb ad Durinck (1992).

For a critical review of censusing methods the reader is referred to Verner (1985). Ralph and Scott's (1981) book contains a wide collection of papers on many aspects of the problems of counting birds, and Wiens (1989) provides abundant illustration of the extent to which methodology is critical to the interpretation of bird counts. A pessimist reading some of this material may conclude that bird counting is so unreliable as to be of limited value. In this book, we take a more positive view. There are many practical reasons for counting birds and it is possible, with adequate care and awareness of method limitations, to produce a study with valuable results.

Counting birds

Among the simplest of aims is to try to discover what species occur at a site, e.g. as a baseline survey of a poorly known site, prior to a more descriptive study. A species list, perhaps with approximations of numbers or abundance categories (common, scarce, rare, etc.) may be surprisingly informative (Chapter 6). Such are the aims and methods of atlas studies (Chapter 10). More elaborate methods might be developed to count particular species (Chapter 8) or deal with large concentrations of birds (Chapter 9). It will almost never be possible to apply any one method to counting all species, as some birds are much more difficult to find and count than others.

Stock-taking (counting exactly how many birds exist in an area), can be appropriate for extensive studies of some species. The methods may be quite crude but are nonetheless valuable, especially if it is feasible to count the whole population over a wide area. Such surveys often require help from a large number of fieldworkers. It is very difficult to judge how accurate the results are because flaws in the methods may be poorly known, i.e. there may be no way of telling if all areas occupied by the species are covered equally thoroughly by all fieldworkers.

For particularly obvious species, features of their abundance and range may be clear enough to survive some deficiencies in the method. Population levels of Grey Herons (Reynolds 1979) and Rooks (Sage and Vernon 1978) have been monitored by fairly simple counts and with useful, even if not totally accurate results. Waders are notoriously difficult to count but are popular subjects for studies both in breeding (Smith 1983) and non-breeding (Prater 1981) habitats. Many wading bird species tend to breed at low densities over very large and remote areas. Distinct breeding populations may winter on the same estuary, or individuals from the same breeding population may winter at distinct sites. Different ages or sexes of birds may winter in different areas. Individuals may even utilise a number of different feeding areas at different times of the year. Ideally, extensive population studies have either to guarantee a uniform effort in all areas, or effort needs to be measured. This means first identifying all the sites a species may occupy, and recording any sites that are not surveyed for one reason or another (Fig. 1.1).

If the study area is a small nature reserve, wood, farm or marsh, then mapping methods (Chapter 3) may be used to survey the numbers and distribution of breeding birds. Mapping methods are very labour intensive, but this may not matter on such small sites. The great advantage of mapping is that the results can be used to identify the approximate location of bird territories. A common use of such maps is to estimate what effect losses of habitat might have on birds. One can show which parts of a plot hold particular species or habitat/species combinations, and then argue that the loss of these areas would have a big impact on the local bird community. Someone who knows the site, i.e. a volunteer fieldworker, will understand the influence of local habitat on the mapped distribution of species. If an outsider has to interpret the bird data, then it is vital that the fieldworker produces a very good habitat map (Fig. 1.2). If a bigger area or a wider comparison of habitats is needed, then a sample of transects or point counts might be more appropriate (Chapters 4 & 5).

Distribution studies

Certain types of study aim not to count birds at all, but merely to specify where they do and do not occur. Distribution maps may illustrate the presence of species in some units of area, i.e. 10 × 10 km squares nationally or 2 × 2 km squares (tetrads) for smaller scale surveys. The general presumption of such an approach is that species' presence can be reliably detected given enough effort. To be of interest, such maps also need to show areas with and without birds, which depends to some extent on the spatial abundance of the species and the size of the units mapped (Fig. 1.3). Coverage needs either to be reasonably uniform or to be measured and reported (Fig. 1.4). Otherwise, the resulting maps will show distribution of observer effort as much as distribution of birds. Distribution studies are described in more detail in Chapter 10.

Figure 1.1
———————————
Population
counts with
varying
effort

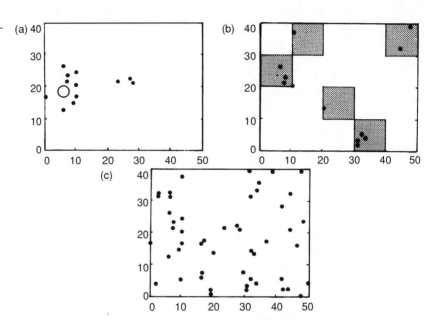

A hypothetical county happens to be a perfect rectangle and the recorder is interested in describing the Magpie population. Three different approaches are tried in succession.

(a) **Map of all Magpie nest areas in the past 5 years:** what does it show? A possibility is that 10 km squares 01 and 02 are good for Magpies which are generally rare in the county. Perhaps there are about 20 pairs in the whole county allowing for a few which have been missed?

(b) **Sample survey:** five randomly selected 10 km squares were fully covered and the 12 nests marked were found. What does this mean? Square 02 now looks less special. Four nests were also found in square 30 where Magpies had not been recorded casually as shown above. The total population for the county can be estimated from this survey as 12 x 20/5 = 48 pairs. These squares could be re-visited in future years to assess population changes.

(c) **A full survey:** this shows that there were actually 50 pairs. Assuming that the method was accurate, this figure is correct.

The sample survey took only a quarter of the effort and produced quite a good answer for the total population but did not, of course, actually locate most of the pairs. The sample survey was good enough to get total numbers but not sufficient if location was also important.

The data from the full survey could be analysed to explain variation of density in terms of habitat.

Example (a) was totally wrong. This was because the ringing group based in the town marked with a circle was very keen on Magpies and submitted all the records. No one else thought Magpies worthy of note!

Figure 1.2

Mapping
census and
habitat map

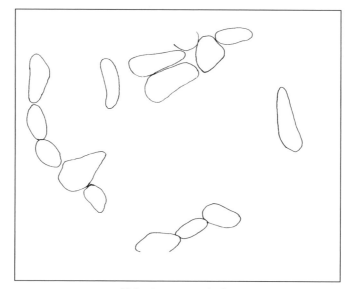

Yellowhammer territories

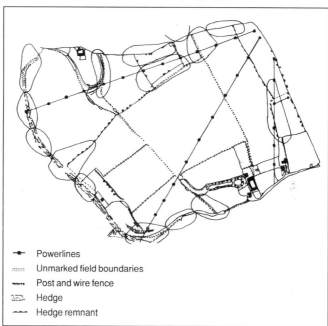

● — Powerlines
⋯⋯ Unmarked field boundaries
+++++ Post and wire fence
⊃≡⊃ Hedge
⌒⌒ Hedge remnant

(a) Map of Yellowhammer territories on a mapped plot (Williamson 1968). It means very little on its own if you happen not to know the site where it was done.

(b) It makes more sense when some key habitat features such as field boundaries have also been mapped in. Yellowhammer seem to prefer territories along hedges or other field boundaries, with few in the centres of fields.

Figure 1.3

Mapping
distributions
at different
levels of
scale

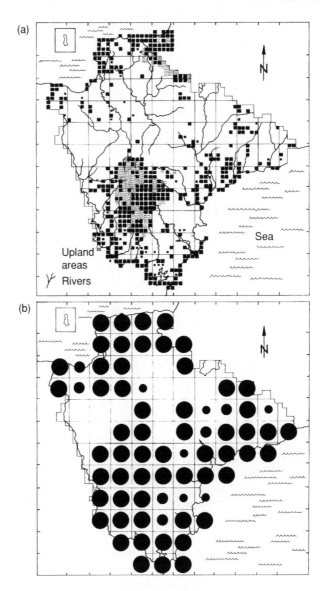

(a) Distribution of Stonechat from the Tetrad Atlas of Breeding Birds in Devon (from Sitters 1988). Stonechats occur especially on the upland areas of Exmoor, Dartmoor and in coastal regions. The survey results are presented in 2 × 2 km squares. A repeat survey will be able to show any distribution changes.

(b) The same data redrawn by 10 × 10 km squares which is the standard used for national atlases in Britain (e.g. Lack 1986). The details of factors influencing distribution have been obscured. It will take a much bigger range change before this scale of work can detect it.

Figure 1.4

Survey
coverage
affecting
results

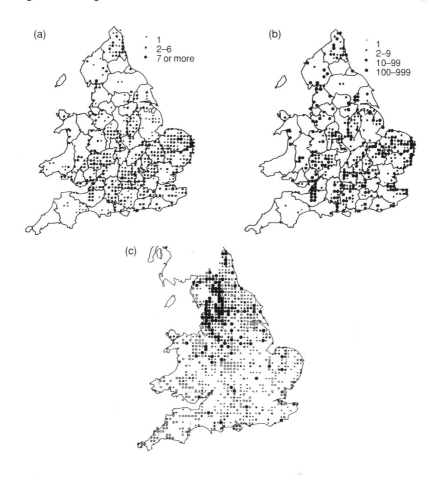

(a) The distribution and numbers of grassland sites surveyed in each 10 km square. All survey sites are included even if no breeding waders were located (from Smith 1983).

(b) The distribution and numbers of breeding Lapwing pairs found on grassland in each 10 km square (from Smith 1983).

(c) The distribution of nesting Lapwings in England and Wales in 1987 (from Shrubb and Lack 1991). The symbols represent numbers in one tetrad in each 10 km square: small dot = 1–4 pairs recorded; medium dot = 5–10 pairs; large dot more than 10 pairs; blank = the tetrad was visited but held no birds; and an open circle = not visited.

What does it mean? It is very difficult to infer much from the bird map (b) because it is influenced by the coverage map (a). The coverage map may represent the distribution of suitable habitats but this is not claimed because coverage was incomplete. The species distribution map (b) does not, in fact, represent the breeding distribution of the Lapwing which reaches its greatest abundance in the north-west in England (c).

Population monitoring

Trends in numbers over time are of particular interest to conservationists. A species may be inherently rare and thus in need of surveillance. It may occupy habitats known to be changing, or perhaps be a candidate for indicating the adverse effects of pesticides or pollution. Populations may fluctuate naturally as a result of the combined effects of weather, predation and other factors that influence reproduction and survival. This year-to-year variation could easily conceal a decline/increase in population for several years if the net annual change in population was actually quite small. Known fluctuations due to weather or natural population dynamics therefore need to be measured with some confidence if they are to be accounted for. The ability to distinguish between natural and man made population changes is an essential attribute of a successful monitoring scheme (Baillie 1990).

Species in which the whole population can be located with reasonable confidence are not so difficult to survey. This includes any species that either exists conspicuously in open habitats, species that exist in specialist habitats that are well known and infrequent, or species whose biology and behaviour is known well enough for them to be located easily. Thus the number of Marsh Harriers breeding in Britain has been well monitored over a great many years (Fig. 1.5). This is because it is a charismatic species and many people have checked the few suitable patches of habitat within the known range each year. Almost complete counts have been made at periodic intervals of birds as diverse as Great Crested Grebes, Peregrines,

Figure 1.5

A complete population survey

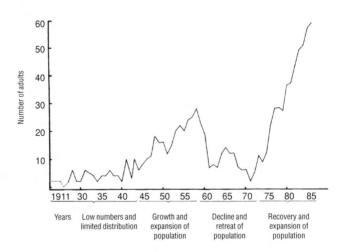

The numbers of breeding adult Marsh Harriers in Britain, 1911–1986 (from Day 1988). Problems with sample coverage and design do not arise in interpreting what this means. The only question is whether the coverage was complete for a study dating back so far. It is believed that it was for such a conspicuous bird which breeds in a restricted range of habitats.

Gannets and Dartford Warblers. In all these examples, valuable results have been obtained in spite of methodological imperfections.

If the species is numerous or widespread, population monitoring may be no less important but complete counting could be complicated. Moreover, these species are often smaller and therefore harder to count. It is relatively easy to tell if a crag has a pair of Peregrines or not. It is substantially harder to tell if a wood has ten pairs of Chaffinches or 20 pairs. If a complete count is out of the question, then a sample must be selected with some care. Random sampling of 10 km squares (or similar units) is possible. Such a system has been used with Mute Swans (Ogilvie 1986) and Wood Warblers (Bibby 1989). Samples of auk breeding colonies (Stowe 1982) on the other hand, have been criticised because their distribution was not random. An important point is that, if sampling is non random or biased (see Chapter 2), it may be less justifiable to infer statistics about the population as a whole. It is thus important to understand just what has been sampled and how it might relate to the whole population in connection with such factors as geography, habitat and species biology.

The most long standing population monitoring scheme in Britain is the Common Birds Census (CBC). This is a mapping method described in more detail in Chapter 3. It is tempting to believe that the published indices for farmland and woodland apply to whole populations (Fig. 1.6). In fact they apply very strictly to the places from which they came and are strongly biased to south and east England (Fuller *et al.* 1985). Internal

Figure 1.6

Trends in
population
derived from
indices

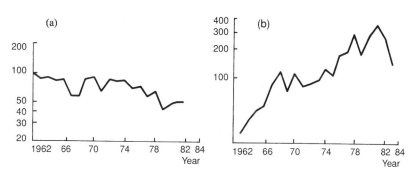

Trends are shown for breeding Lapwing populations in areas of England and Wales dominated by cereals (a) and sheep rearing (b) (from O'Connor and Shrubb 1986). The trends are derived by estimating the proportional population changes from one year to the next on plots that were counted by the same observer and methods in both years.

The implication that these are general results seems surprising from casual observation, which suggests that Lapwings have disappeared very widely from cereal areas but have also declined markedly in sheep-dominated areas such as Wales. The generality of the results depends on how the study plots were chosen and where they were located. This information was not given. Without it, it is not possible to say exactly what these plots mean.

Figure 1.7

Population
trends from
studies
using paired
plots

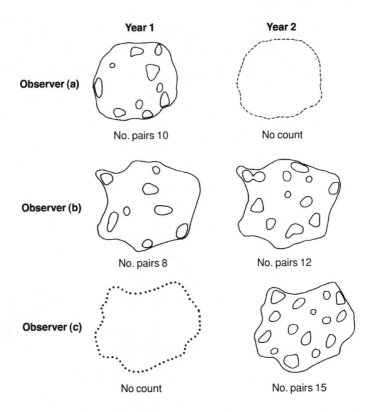

Three mapping plots (a–c) in one county are studied in 2 years. All plots are 25 ha in extent. Hypothetical results show territories of Wood Warblers. What does it mean?

The mean density of Wood Warblers was 36 pairs per km^2 in year 1 and 54 pairs per km^2 in year 2. There was a 50% increase from year 1 to year 2 as shown by mean density on all the plots studied in each year. As it happens, the one plot studied in both years also showed a 50% increase. So might we infer that Wood Warblers increased in numbers by 50% in the county from year 1 to year 2 and that they generally occur at densities of 30–60 pairs per km^2?

Not necessarily. The observers chose their study areas and picked interesting-looking woods, not typical of the county where few of the woods actually support Wood Warblers at all. So no reliable general estimate of Wood Warbler density in the county's woods can be made. Half of the woods in the county blew down in a gale in winter of years 1 and 2, which is why observer (a) gave up his plot in year 2. The total number of Wood Warblers in the county actually went down by 25% between the two years. This is not represented in paired plot studies of this kind if observers stop counting on plots that have become unsuitable.

This illustration is deliberately exaggerated to make its points. It would be obvious that half the woods had blown down. It is easy to see that more subtle changes could be overlooked and the results could mislead.

checks are made to ensure that within this area, the plots are representative of the range of agricultural practice and observers do not give up squares when they become less interesting. It is easy to see how a non-random sample could fail to show up the effects of habitat deterioration for such a reason. One could imagine observers giving up when their farm is converted into a wheat prairie, while a new starter might be tempted to select somewhere with mixed farming and hedgerows (Fig. 1.7).

More recently, CBC has been replaced as a population monitoring exercise, by Breeding Bird Survey (BBS). BBS is based on a line transect method. It also uses randomly selected 1-km UK grid squares as the basis of its samples, but it weights the results according to observer effort in different regions of the country.

Another problem in population monitoring is that the methods should be comparable from year to year. To compare annual counts, the estimated population of a species does not have to be the same as the actual population. As long as the degree of error in measuring the number of birds remains constant each year, changes in the number of birds recorded can be attributed to changes in population (Chapter 2). For example, provided an observer with poor hearing detects twice as many Goldcrests after numbers double, it does not matter if he/she only recorded two and four pairs when in fact there were ten and 20.

Assessment of habitats for birds

When combined with habitat data, bird studies have a number of different applications. For example, they can be used to predict the effects of changing land-use. They may be used in Environmental Impact Assessments (EIAs) or in the management of areas such as nature reserves.

Studies that attempt to derive conclusions about habitats may need to be more detailed. That is, a greater effort may be needed if the results from any one habitat are accurate enough to compare with the results from another (see Chapter 10). Differences or similarities in the bird counts between habitats can be confounded by a number of factors. In woodland, these could include size/age of trees, understorey density or canopy cover. In more open habitats more species and numbers of birds may be seen. If the area is hilly, then the number of birds may be affected by the direction that the slope faces. To deal with the effect of such variables, it may be necessary to collect a better quality of bird data. This depends on first identifying particular confounding variables and then adjusting the methodology to take account of them (Table 1.1).

Another reason for increasing effort is that a small sample will find few records of scarce species. It may be that these are more confined to rare habitats and therefore of more interest to the study. In coppice areas in woods, for example, the plots cut are rarely larger than a hectare, so most species are absent or represented by a single pair whose territory boundaries may well be wider than the plot. If the sample area is too small to show any meaningful results, the study needs redesigning before any data are

Table 1.1

A hypothetical study designed to show the effect of age of trees on bird communities in conifer forests produced the following results from four equal-sized mapped plots

Plot	A	B	C	D
Age (years)	2	5	10	40
Tree Pipit	0	6	1	18
Nightjar	0	0	3	0
Great Spotted Woodpecker	1	0	0	0

It is impossible to tell what the data mean! The plots differed in many attributes, other than age, which might have caused the differences. Plot D was surrounded by a large clear-fell which attracted many Tree Pipits to its edge. Plot C was the only one on a sandy site which is why it had Nightjars, the rest were on clay. Plot A was adjacent to an oakwood which is where the Great Spotted Woodpecker nested. The possibilities for explaining the results are endless. The inference that Tree Pipits favour older stands and Great Spotted Woodpeckers prefer very young trees would be wrong.

What could be done to improve the study?

(1) Match the plots more carefully to reduce the number of factors in which they differ.
(2) Spread the survey more widely so there can be replicates for areas of different age. Using point counts or transects would allow this.
(3) Study a small number of plots for several years, though 40 years would be a long time to wait for the results.

collected. Habitats may be studied using a mapping census to compare distribution of birds within a plot, in relation to vegetation features (see Chapter 3). Efficient methods such as point or transect counts, however, may be preferred when a large sample is required (Chapters 4 & 5).

Choosing a method of recording habitats varies in the same way as deciding on a method to record bird numbers. How detailed a method is depends on the particular aim of the study, its time span and the level of precision needed to obtain an adequate result. Because habitat survey underpins many bird studies, it is of great importance and is discussed in detail in Chapter 11. Even after habitat data have been collected, there are difficult analytical problems. Computers and elaborate multivariate statistical methods may be required. This area is largely beyond the scope of this book but a brief discussion is provided in Chapter 11.

Management experiments

Experiments to test hypotheses about birds and habitats are often limited by the large scale of vegetation manipulation required. More often observation trials are conducted by taking advantage of a planned change, be it intended to encourage wildlife on a nature reserve or prejudicial as in a so-called farm improvement scheme. The before and after measurements may be made in successive years. But how can you tell whether a change in a particular species was due to the experimental factor or to an extraneous factor such as climate?

First, it is essential to know what would have changed on the plot if the proposed alteration had not occurred (Figure 1.8). The easiest way to do

Figure 1.8

The use of an experimental control

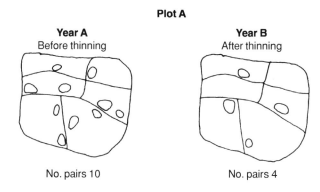

Plot A

Year A
Before thinning

Year B
After thinning

No. pairs 10   No. pairs 4

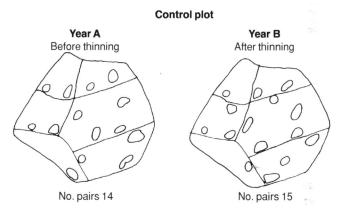

Control plot

Year A
Before thinning

Year B
After thinning

No. pairs 14   No. pairs 15

The figure shows a hypothetical experiment to test the effects of thinning of conifer forest plantations on populations of Wrens. One plot (A) was studied in the years before and after thinning.

What does the decline in numbers mean? It is not clear. The snowfall in December was the heaviest in living memory and Wren numbers may have suffered even if the forest had not been thinned. But the national Common Birds Census index went up by 8% from year A to year B. Unfortunately this is not very helpful as the study area is in northern Scotland and the CBC index is biased to southern England which did not have much snow.

What could be done to improve the study? Include a control plot without any vegetation change. This will measure the effects of other factors such as weather. In the control plot in this study, the numbers of Wrens did not decline.

Conclusion: Wrens did not suffer severe mortality in the great blizzard (it thawed very fast). The decline in numbers was therefore probably due to the thinning, but without replication it is impossible to say how repeatable this result might be.

this is to have two identical plots: one on which the proposed change will take place, and one that remains untouched. This latter example is known as the control. Obviously, the same level of census effort is required at each plot before and after the alteration has taken place. The results then just need to be compared with one another to see if a significant change has occurred between the two plots. Ideally the experimental plot and the control should exist as near to one another as possible, on the same soil type, experience the same weather conditions and both be studied in a single season. Unfortunately this is not always possible, especially if there are a large number of plots or the researcher depends on surveying private fields under different management regimes on different farms.

Conclusions

Throughout the rest of the book, we have attempted to illustrate the use of particular methods to meet particular aims. Our examples are necessarily very selective. Table 1.2 indicates a range of recent studies classified according to their methods and purposes. This shows where to look in the book for further discussion and also offers some pointers for additional reading.

This chapter has outlined a range of possible objectives for a census study and promoted the idea that particular methods are suited to particular aims. The rest of the book describes the variety of such methods, available both for counting birds and describing habitats. Figure 1.9 provides a guide to the costs of carrying out certain survey methods.

Effort spent matching methods to aims will be more than repaid if the survey is robust and efficient. So in order to specify exactly what the aim is, ask the following questions:

- What scale of result is needed? i.e. Is it necessary to know exactly how numerous a particular bird is or will an index suffice?
- Must it apply to a small area and one habitat type or to many habitats and a large area?
- The number of study plots will largely define the scale of the work but then will the results meet the aim of the project?
- Subsequently, is a complete survey needed or is the proposed scale so large that sample plots will have to be used?
- If samples are to be used, how will they be selected?
- What accuracy and precision of results are needed (see Chapter 2) and how will they be achieved?

It is a common misconception that methods of analysis should be considered *post hoc*. This is by no means true! Thought should be given to the methods of analysis before a census method is chosen. The method of analysis has to be capable of producing results that will illustrate the answer to your questions about bird numbers. In turn, data have to be collected in a way that is appropriate to that very method of analysis.

Figure 1.9

Time-costs
using a range
of different
methods of
censusing

Before developing a survey methodology, it is useful to give some thought to the time it will take to carry out using any particular method. What follows is a comparison of the time taken to carry out several methods of surveying to illustrate one reason why they are used in different situations.

Line transects and constant effort searches
When surveying large areas, sampling methods are the most appropriate. For instance, a pilot survey of breeding Golden Plover using line transect distance sampling (a) was undertaken on a 45.4 km² area of upland moorland (Buckland et al. 1998). Birds were recorded along with a description of their breeding status. The detection function for Golden Plovers was modelled and a density calculated as number of pairs/km². The results agreed with estimates from previous surveys using different methods.

The time taken to conduct this survey is compared to a constant effort method of surveying upland breeding waders (b) where all areas of the site are covered to the nearest 125 m (Brown & Shepherd 1993). Buckland et al. (1998) discovered that about half of all birds were overlooked on the line transect survey. Therefore, researchers using constant effort methods must be aware that there is a risk of missing quite a large proportion of the birds. The method is also considerably less efficient but may be useful for surveying small areas when it is difficult to get a reasonable sample of line transects to estimate densities accurately.

(a) The effective width of the transect* (calculated from a significant decline in detectability) was about 150 m. 65 km of transects were surveyed by 4 people in 30 hours.

Effective strip width*	150 m
Length of transects	65 km
Site area	45.4 km²
Sample area	9.75 km²
Number of observers	4
Number of visits	1
Number of man hours fieldwork	30
Man hours/km²	**0.17**

(b) For an equivalent area of 45.4 km²

Number of observers	4
Number of visits	2
Number of hours per km²	1.2
Number of man hours fieldwork	109
Man hours/km²	**2.4**

Point transects
For various logistical reasons, it is unlikely that point counts would be used on an upland wader survey (see Chapter 5). For example, if we assume that a 10 minute observation period is required and that we cover the same sample area as the line transect survey and the effective 'radius' is equal to the effective 'strip width', it would take about 2.5 man hours to survey each square kilometre. This method is therefore much less efficient than line transects. Nonetheless, in woodland surveys it is used as an alternative to line transect surveys owing to numerous logistical restrictions (see Chapter 4).

Territory mapping
Territory mapping is one of the least efficient methods of surveying birds, but serves a practical purpose for surveying small sites, particularly when breeding activity and distributional data are required (see Chapter 3). It would never be practical to apply the method to a large upland wader survey. For example, according to the BTO Common Bird Census (Marchant 1983) it takes approximately 3.5 hours to thoroughly map an area of 70 ha of farmland. This figure could easily be lower on open moorland, although the terrain is harder to traverse. Less time would also be spent if only Golden Plover was recorded so the figure may be closer to 80 ha, or 0.4 km²/h. Assuming we carry out ten visits, to cover an equivalent area as in the line transect example would require 22 hours per square kilometre.

Summary and points to consider

1. What is the purpose of the study?
Is the aim a reasonable one?
What can be learned from reading about previous studies?
What scale of precision is required?
Which and how many species need to be included?
Are approximations, indices or absolute numbers needed?
What sort of sample size is needed?

2. What are the field methods?
What basic methods (maps, point counts, look-see, etc.) should be used?
How many plots/routes/points should be used?
How are sample areas to be chosen?
How much will it cost (time or money)?
Are the observers skilled or how will they be trained?
What are the likely sources of bias?
What steps will be taken to deal with bias?
How will the results be recorded – design of forms, etc.?

3. Do the methods suit the purpose?
Are the methods sufficient but not excessive?
Do any other variables need to be measured?

4. Sample sizes?
Are the sample sizes sufficient?
Has effort for scarce species been increased since they are more difficult to detect or have a lower probability of detection, especially important in EIA work?
Are there enough data points to relate to habitat information, where this is required?

5. How will the data be analysed?

Table 1.2

Examples of
key studies

GENERAL REFERENCES

General methods	Bibby et al. (1998)	Expedition field techniques (Bird surveys)
	Gilbert et al. (1998)	Bird Monitoring Methods: a manual of techniques for key UK species
	Komdeur et al. (1992)	Manual for Aeroplane and Ship Surveys of Waterfowl and Seabirds
	Walsh et al. (1995)	Seabird Monitoring Handbook for Britain and Ireland
	Davis (1984)	Methods of censusing birds in Australia
	Dunn et al. (1995)	Monitoring birds in Canada
	Ralph et al. (1993)	USA bird census methods
	Cranswick et al. (1997)	UK Wetland Bird Survey (WeBS)
	Sutherland (1996)	Ecological Census Techniques: a handbook
Conservation status & population	JNCC (1996)	Birds of Conservation Importance (UK)
	Gibbons et al. (1996)	Bird Species of Conservation Concern (UK)
	Collar et al. (1994)	Birds to Watch 2: The World List of Threatened Birds
	Tucker & Heath (1994)	Birds in Europe: their conservation status
	Stone et al. (1997)	Population estimates of birds in Britain and in the United Kingdom
Statistics	Fowler & Cohen (1990)	Basic Statistics for Field Biologists
	Sauer & Droege (1990)	Survey designs and statistical methods used in USA
	Zar (1984)	Biostatistical Analysis

GENERIC METHODS

Absolute counts	Day (1988)	Marsh Harriers in the UK
	Marquiss (1989)	Grey Herons in the UK
	Green (1995)	Corncrake in the UK
Territory mapping	Marchant (1983)	Common Bird Census in the UK
	Shrubb & Lack (1991)	Numbers and distribution of Lapwings
	Stowe & Hudson (1990)	Status and distribution of Corncrake in Britain
	Taylor (1982)	Waterways Bird Survey in the UK
	Terborgh et al. (1990)	Territory mapping in the tropics
	Moyer (1993)	Passerines in Tanzanian forests

Distance Sampling	Buckland et al. (1993)	Distance Sampling book
	Thomas et al. (1998)	Distance 3.5 (Software and user's manual)
Line transects	Buckland et al. (1998)	Monitor selected species on designated sites
	Gregory & Baillie (1998)	Large-scale habitat use of British birds
	Gregory et al. (1997)	The Breeding Bird Survey 1995–1996
	Lindén et al. (1996)	Wildlife triangle scheme in Finland
	Tasker et al. (1984)	Seabirds at sea
	Viñuela (1997)	Road transects for Red Kite in Spain
Point counts	Bibby & Charlton (1990)	The Azores Bullfinch
	Desante (1981)	Variable Circular Plot (VCP) method
	Marsden et al. (1997)	Using VCP in a tropical forest
	Reynolds et al. (1980)	VCP method
	Whitman et al. (1997)	A comparison of mist-netting and point counts
	Marsden (1999)	Parrot and hornbills in Indonesia
Capture and marking	Boddy (1993)	Population studies on Whitethroat
	du Feu et al. (1983)	du Feu method
	Hestbeck & Malecki (1989)	Lincoln Index
	Peach et al. (1998)	UK constant effort ringing scheme
	Whitman et al. (1997)	A comparison of mist-netting and point counts
Radio tracking	Green (1984)	Grey & Red-legged Partridge ecology
	Hill & Robertson (1988)	Pheasant ecology
	Hirons & Johnson (1987)	Woodcock ecology
	Jouventin & Weimerskirch (1990)	Satellite tracking of Wandering Albatross
Migration counts	Bildstein & Zalles (1995)	Raptor migration counts
	Lowery & Newman (1956)	USA autumn bird migration
	Pyle et al. (1994)	Nocturnal migrants in Farallon Island, California
Vocalisation studies	Redpath (1994)	Censusing Tawny Owls using imitation calls
	Marion et al. (1981)	Using playback recordings to sample elusive birds
	Gilbert et al. (1994)	Vocal individuality as a census tool
	Ratcliffe et al. (1998)	Playback census for Storm Petrel
	Haug & Diduik (1993)	Playback for censusing Burrowing Owl
	Mosher et al. (1990)	Playback of woodland raptors in USA
Look-see	BTO (1991)	The fourth BTO Peregrine Survey 1991
	Clarke & Watson (1990)	Hen harrier winter roost survey in the UK
	Cranswick et al. (1995–96)	UK Wetland Bird Survey (WeBS)
	Watson et al. (1989)	Golden Eagle survey
	Crooke et al. (1992)	Slavonian Grebe survey
Randomised look-see	Rebecca & Bainbridge (1998)	Survey of Merlin in Britain
	Underhill et al. (1998)	Survey of Common Scoter in Britain and Ireland
	Hancock et al. (1997)	Greenshank in the UK

Distribution	Sharrock (1976)	Atlas of Breeding Birds in Britain and Ireland
	Gibbons et al. (1993)	New Atlas of Breeding Birds in Britain and Ireland
	Lack (1986)	The Atlas of Wintering Birds in Britain and Ireland
	Holloway (1996)	Historical Atlas of British birds
	Price et al. (1995)	North American summer bird Atlas
	Harrison et al. (1997)	The Atlas of Southern African Birds
	Hagemeijer et al. (1997)	The EBCC Atlas of European Breeding Birds
	Bibby & Hill (1987)	Distribution of Fuerteventura Stonechat
Species richness & diversity	Bibby et al. (1992)	Birds as indicators of biodiversity
	Stattersfield et al. (1998)	Endemic bird areas
	Williams et al. (1996)	Bird distribution patterns in the UK
Timed searches/ counts	Brown & Shepherd (1993)	Upland wader surveys
	McKinnon & Phillips (1993)	20-species list method
	Pomeroy (1992)	Timed Species Count
Radar	Burger (1997)	Numbers of Marbled Murrelets
	Liechti et al. (1995)	Bird migration in Israel
Satellite data	Russell et al. (1998)	Roosts of Purple Martins
	Miller &Conroy (1990)	Wintering habitat of Kirkland's Warbler
	Lavers & Haines-Young (1996)	Dunlin distribution in northern Scotland
	Pienkowski et al. (1993)	Bird census in remote areas

2

Census Errors

Introduction

The numbers of birds in a particular place or the average density of birds in a habitat for one season has a precise value – the true value, which is generally unknown to us. Because some birds are highly mobile, elusive, widely distributed or migratory, it is not always possible to carry out a complete count (like you might count heads on a bus for example), so we conduct experiments to try to estimate what the true value is. Sometimes these experiments are 'sample' experiments and only look at a proportion of the area that the species occupies. This chapter discusses considerations that apply when designing an experiment so that the estimate obtained from a sample is as accurate and precise as possible.

The difference between our estimate and the true value is called the error (the word being used in a statistical sense rather than in its common meaning of a mistake) and we can measure it to determine how good our estimate is. Therefore, understanding errors is of great importance to designing and interpreting a study and whatever else, they should not be ignored when interpreting results. There are two sources of error: normal variation and bias. Results with minimal amounts of each are known as precise and accurate (or unbiased) (Fig. 2.1). The concepts of precision and accuracy are very important and must always be considered when designing a bird-counting study.

Imagine that we want to measure the average density of Skylarks in a relatively uniform area of arable farmland. Assume that we have some way of telling exactly how many pairs there are in a small plot within this farmland. The density on this plot might be quite similar to the overall farmland density but it is unlikely that it will be exactly the same. If we take the average density from several small plots, the result will probably be nearer the actual density. If we take 100 plots, the result will probably be even closer to the true density. A frequency histogram of the results from each of the 100 trials would have an approximately symmetrical shape, with a hump in the middle (most of the results being grouped around the true value). With an infinite number of plots, this could be represented by a smooth curve: the Normal distribution. The Normal distribution is of widespread occurrence and the basis of a large part of statistics. Fowler and Cohen (1986, 1990) provide a more rigorous introduction to Normal

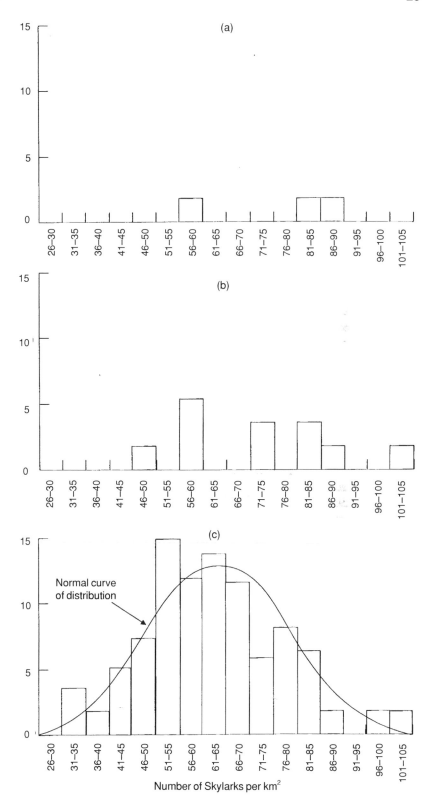

Figure 2.1

The normal curve of distribution

Figure 2.1

―――――――――

(continued)

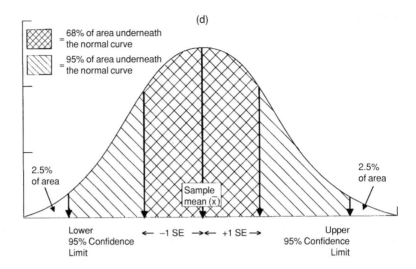

The figure shows hypothetical densities of Skylarks from small sample plots in a large uniform area of habitat. The true density is 70 birds per km², SE = 20.

(a) Three study plots show results scattered about the truth (which is unknown to the observer; sample mean 81.6 birds per km²).

(b) The mean of 10 results is quite close to the truth (sample mean = 75.0 birds per km²).

(c) One hundred plots would take a huge effort to cover but give a good impression of the mean density and its variation from plot to plot (sample mean = 71.2 birds per km²).

(d) An infinite number of plots may result in a smooth curve of distribution, the Normal distribution. The hatched areas indicate the proportion of plots whose densities fall symmetrically, ± 1 Standard Error (equivalent to 68%) around the mean, or within 95% of records that fall symmetrically around the mean – the 95% confidence interval (CI). The densities along the **x**-axis (indicated by the vertical arrows) are the boundaries of these regions.

curves, with examples drawn from the bird world. In summary, a single sample is very unlikely to be exactly representative. As more samples are added, their average becomes closer to the true value and the result becomes more precise (Fig. 2.2).

Precision

Precision is a measure of 'normal variation' e.g. how much the results deviate from the true value (see Fig. 2.1). If most of the results are closely grouped around a mean, then the data can be said to be precise. If there is a lot of variation in the results, the opposite is true. There is no way of

Figure 2.2

Precision
increases
with sample
size

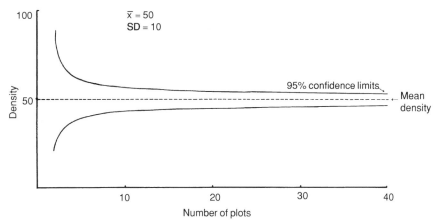

The figure shows the relationship between the number of plots sampled and the confidence which can be put on the result for a particular hypothetical population.

telling in advance how varied the data will be, without carrying out a pilot study or referring to other similar projects. Therefore, it is vital to measure this variation and design an experiment that produces results that are as precise as possible.

One source of statistical error that can result in imprecision is measuring or sampling error. Measuring errors may also be distributed in a normal curve. For example, if 100 people used the same ruler to measure your height, there would be some variation in the results but they would probably be closely grouped about the true value.

If it can be proved that variation in density from one small plot to another is little to do with measuring errors, then it is a property of the population being studied. In our Skylark example (Fig. 2.4), the error causing imprecision arose from the fact that we surveyed only a small part of the total area and that the birds were not uniformly distributed. Such spatial variation might have been due to chance, or because the habitat was not as uniform as first assumed. If the density of a species varies considerably from place to place, the area sampled may need to be larger in order to include a whole cross section of densities. Making the plots bigger would be one way of increasing the sample size. The correct response, however, is not to put a lot of effort into one huge plot and get a single answer but to count many smaller plots. Although a large sample of small plots increases the chance of recording extremely high and low values – and increases the variation in density – it also increases the proportion of records near to the mean. When a large percentage of the results are distributed around the mean, precision can be measured by identifying the limits of a proportion of these results, i.e. a proportion of the area under the normal curve of distribution (Fig. 2.2).

In order to test how good our sample mean ($\bar{x}$) is as an estimate of the true mean, we need to calculate the standard deviation (s) and the standard error (SE) of our estimate. The formula to calculate SE depends on the number of samples (n) and a measure of the average variation of each result (x) from the sample mean. This measure is called the standard deviation (s). Equations for calculating mean, standard deviation and standard error are shown below.

$$\text{Sample mean } (\bar{x}) = \frac{\Sigma\,(x)}{n}$$

$$\text{Standard deviation } (s) = \sqrt{\frac{\Sigma\,(x - \bar{x})}{n - 1}}$$

$$\text{Standard Error (SE)} = \frac{s}{\sqrt{n}}$$

The probability that the true mean falls within one standard error of the sample mean is always 68% (i.e. 68% of all samples drawn at random from a normal distribution will fall within ±1 SE of the mean). However, we generally want to be more precise about the range of values which contain the mean, so we employ a more stringent measure of precision, the common conventional value of 95%. In the case of normally distributed data with a large (theoretically infinite) number of samples, 95% confidence limits can be calculated simply by multiplying the standard error by ±1.96. For sample sizes of less than 30, it is generally recommended that a better multiplier is looked up in the table of the Student's t-distribution (Fowler and Cohen 1986, 1990; Sutherland 1996). As more samples are added, so the range between the confidence limits will tend to contract (as more results are grouped around the mean). Therefore, these methods formally measure the obvious concept that the larger the sample size, the closer the average will be to the true value (Fig. 2.3). Some surveys therefore require much more data collection than others to obtain the same precision in their estimates, according to the variability of the data.

It is worth noting at this point that not all bird data are distributed in a normal curve. In fact, it is very common for biological data to be skewed towards zero, as there tend to be very many plots where no, or very few birds are recorded. The same rules concerning error are applicable to non-normal data, but there are special statistics – non-parametric statistics – to test these errors.

Precision can therefore be measured and increased by taking more samples, although this is more time consuming. Unfortunately, precision only increases in proportion to the square root of sample size (e.g. to double the precision obtained from ten samples requires $10 \times 2^2 = 40$ samples). To double it again would require $40 \times 2^2 = 160$. Clearly, at this scale of multiplication the effort required to obtain precise results rapidly becomes unrealistic. The precision that can be obtained has to be balanced against the time available. It also has to be considered in relation to the

Figure 2.3

Random,
stratified and
systematic
plots

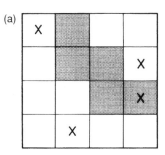 (a)

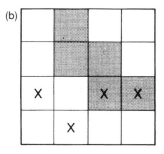 (b)

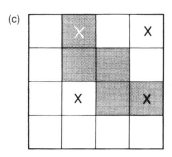 (c)

(a) The area to be studied is divided into square blocks on a map. Plots are selected at random to ensure that the results can be taken to be generally representative of the whole area under study. By chance, three plots fall in one habitat and only one in the other.

(b) The whole area is again divided into square blocks on a map. These can be attributed to each of two habitats which are known or thought to differ in their suitability for the study bird. Two randomly chosen plots are picked within each of the two categories of habitat.

(c) Plots are arranged in a pattern where they are exactly one whole square away from each other. The white cross represents the starting point, which was chosen at random from all 16 possible plots.

nature of the question being asked, e.g. is a density estimate with 50% confidence limits good enough, or must it be better? You cannot conclude that the densities at two sites are different if their confidence limits overlap. The precision required therefore depends on how large a difference is expected. If the two densities are actually very different (this can be identified by carrying out a pilot survey), then fairly imprecise estimates will confirm it. If they are only slightly different, then a lot of work will be required to get sufficient precision to prove that fact. Best practice is to predict in advance roughly how large the difference will be, by conducting a pilot study or by reading the results of other similar work. Once the likely variability of results is known, there are statistical methods available to estimate how many samples are required. This is called Power Analysis. If many more samples than could possibly be collected are required, then it is pointless to continue; the study needs redesigning and a great deal of wasted effort can be saved. Similarly (though less commonly) it is possible to waste time collecting more than enough data for the intended purpose. In this case, time could be more effectively used in developing other aspects of the study.

In the Skylark example we are using, a critical feature is that we are trying to estimate the density of birds in some fixed area which is larger than we could survey in full. If the whole site were surveyed as a single plot, then it would not be possible to analyse the variation in data and compare the results to other similar surveys elsewhere. It would merely say how many birds were on the area concerned.

As already explained, it is ideal to replicate the survey by sampling many small plots. It is important to understand that replication involves counting different birds in different plots – the results from each plot are 'independent'. Increasing the numbers of birds counted by making more visits to the same site is not the same thing. In truth, many bird census studies do not measure their precision at all, and even a pragmatic aim of measuring it and trying to get within about 25% of the true value would be an improvement. If a survey uses only two replicate plots and the results are widely different, it is difficult to draw any conclusions about the mean density of birds. Obviously, similar results from just two plots could be due to chance. Even this low level of replication, however, would well repay the extra effort in improving the confidence with which the results can be regarded.

How though, should the areas be selected? There are three main ways of selecting survey sites: random, stratified and systematic (see Fig. 2.4). The critical feature of randomness is that each plot chosen should have equal probability of falling anywhere within the study area. Its location should not in any way be influenced by prior knowledge which might bear on the expected number of birds. This requirement is essential to ensure that the resulting answers are representative of the study area without bias (see below).

Randomly located plots are picked from a numbered list of all possible plots that could be surveyed, using random numbers which are readily generated by computer, or looked up in a table. They should not be picked by eye in any way such as sticking pins into a map. The human eye is very bad at picking a set of random points. Still less should they be selected by looking around and trying to pick a range of places that seem to look average for the area. An increasingly common practice is to use the national grid as a framework and to sample randomly selected 10-km or smaller squares. This makes picking random squares easy to do and is a useful practice in study design.

Stratified sampling may be used to select plots in for example, different habitats. If the abundance of a bird varies considerably from one habitat to the next, the results from separate habitats may have significantly different levels of data variation. The efficiency of a project can be improved by increasing the number of samples in areas where variation is highest (thus achieving more precise estimates) and minimising the number of samples in areas where variation is small (where the result from a plot is almost

Figure 2.4

A biased
sampling
method

(a)

(b)

The relative density of Skylarks is measured at a sample plot by counting all the birds singing during a 10-minute observation period.

(a) A count of all the birds in the sky showed four birds, even though there were actually seven present (three on the ground).

(b) A count of singing birds showed one bird and again there were really seven present. In both cases the number of birds recorded is less than the true number. Census results will almost always be less than the real number of birds present. The result may still, however, be proportional to the true number: if there are twice as many birds in a plot then twice as many might be expected to be counted.

always close to the true value). An example of this approach is given in Chapter 8, (Fig. 8.4) and further explained in Sutherland (1996). Another advantage is that it is possible to sample a larger proportion of infrequent habitats that may be missed by a random or systematic selection. A pooled estimate of abundance for the whole site can be calculated if the proportion of habitats in the study area is known. In the Skylark example we perhaps ought to try to measure densities in different crops, taking some plots in sugar beet and others in wheat. The same effort might then obtain more precise estimates for individual crops than for the wider more mixed area. Strata should be chosen using knowledge of the distribution and abundance of the various habitats, but the number to be chosen in each will depend on the objectives of the enquiry. Sample locations within strata should be selected randomly as detailed above. The habitat variation may not be sharply split like crops but be continuous like the scrub cover on some grassland plots. In this case, plots could be spread along the range of variation, the scrub cover measured and the analysis performed by other methods such as regression (Fowler and Cohen 1986, 1990).

Systematic sampling is a relaxation of the laws of random sampling. However unlikely, it is conceivable that all plots from a random sample could fall adjacent to one another, or in a single habitat. Systematic sampling involves arranging plots in a pattern but with a random starting point. It is therefore possible that these samples may be biased towards a particular environmental or biological feature (e.g. if sample points coincide with a periodic feature such as a hummock-hollow topography). Care must be taken to ensure that plots are not influenced by some external feature. For example, the Finnish wildlife triangle scheme (described in Chapter 4) has designed a survey that avoids non-random bias introduced by the common northwest–southeast arrangement of glacial valleys.

Accuracy

Accuracy is a measure of how near the estimated (or sample) mean is to the true mean (see Fig. 2.1). Results that are inaccurate are said to be biased. Bias always occurs in a certain direction (e.g. each result will tend to be consistently higher or lower than the true mean). For example, if you do a population estimate of birds and survey mainly in high wind, the results will be lower than the true value because birds are difficult to see and hear in windy conditions.

Bias is systematic in the sense that it cannot be made to disappear with a large sample. In discussing the Skylark example with respect to precision we postulated a perfect census method with no bias. Imagine that in practice we wanted a reasonable number of samples, so we chose a very quick fieldwork method. This entailed standing in the middle of a plot and counting the maximum number of birds that were singing at any one time (Fig. 2.5). Although the count may be larger in a place with more Skylarks, it is unlikely that all the birds breeding in the plot will be singing

Figure 2.5

Precision
and bias

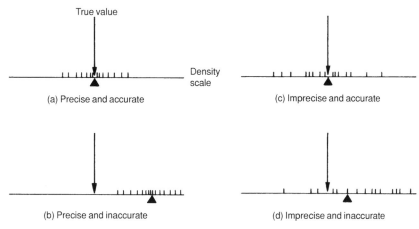

Precision and bias vary independently so either can be high or low in a particular study. The results for 15 different plots are shown in relation to the
unknown true density.

(a) Precise and accurate. The results are closely spaced about the true value.
This is the ideal situation and is probably rarely met in practice.

(b) Precise and inaccurate. The results are closely spaced but their average
deviates from the true value. Since the true value is unknown to the observer,
this result cannot readily be recognised as different from that in (a).

(c) Imprecise and accurate. The results are spread rather widely about the
true value.

(d) Imprecise and inaccurate. The results are spread widely and their average
deviates from the true value. Again, since the true value is unknown to the
observer, this result cannot readily be recognised as different from that in (c).
Some cynics would say that this pattern is the most common result obtained in
bird counts.

simultaneously. Thus, unless birds from elsewhere are flying in and being
recorded, the count will tend to fall short of the true value. The results of
this study would be said to be biased or inaccurate.

Bias is inevitable in almost any realistic bird census method and in
general we know neither how large it is nor in what 'direction' it lies. For
example, in the Skylark example (above), it is even possible that the results
might be inflated if single birds are mobile enough to be recorded on
several different plots. Bias might arise from a combination of several
sources, some of which exaggerate the result while others lower it. It is
common, but not at all desirable, to list several possible sources of bias and
suggest that they may cancel each other out. On the other hand, increasing
awareness of countless sources of bias has made recent bird census
methods much more complicated. If, as is generally the case, accurate
results are very difficult or nearly impossible to obtain, the perfectionist
can argue that it is safest to give up.

It is, however, possible to take a more pragmatic view. This entails being on the look out for bias both in the field and in the literature, with which one might be comparing results. Secondly, though much harder to deal with than random error, bias is not totally intractable. There are three kinds of action available. Firstly, if the likely sources are anticipated, steps can be taken to minimise bias for the particular project, i.e. only surveying in good weather. Secondly, with careful design it is possible to sidestep the problem at the analytical stage by confining comparisons to samples that have the same bias. Finally, it is possible, though often very difficult, to measure bias.

Bias is minimised by recognising its sources (see below) and not allowing them to creep into the study design. Counts may be restricted to a fixed season and time of day. Observers can be trained to reach similar levels of competence. Work can be suspended if the wind is too strong. The counting methods are standardised so that point counts last for a fixed period or transects are walked at a fixed speed. Within a single study, the sources of bias that cannot be totally removed should be spread around fairly. If for instance there are ten plots each to be visited eight times, the order of visits should be designed so that the seasonal spread of visits to each is similar. On mapping plots for example, the direction of walking should be altered across visits. Otherwise one area will tend to be walked at a different time of day from another and spurious differences in bird distribution might be added. The Common Birds Census (Chapter 3) evades observer bias by making year-to-year comparisons only where the same observers cover the same plots in a pair of years. Because the purpose of an experiment is to isolate the factor(s) of interest from other influential variables, avoidance of bias is a prerequisite for all experimental designs.

Bias can only be eliminated *post hoc* if an accurate measure of its true effect can be ascertained. In practice this is extremely difficult. For a single species, it might be done by use of marked birds (Chapter 7), or by nest-finding and mapping on a small subset of study areas (Chapter 3). If different observers come up with different answers then the possible scale of bias can be guessed, but not perfectly measured unless an unbiased estimate has been obtained as a baseline. It is easier to correct for relative bias between observers. Only then is there a possibility that different answers represent a true difference between samples. For a community of birds, an unbiased estimate is even more difficult. Bias associated with vegetation density is a factor in the decline in the detection of a species with distance from the observer. This curve can be plotted using existing software (Chapters 4 & 5). The area under the curve relates to the probability of detecting the bird, so adjustments can be made to remove the effect of variables on detection by estimating the number of birds missed. In this way, if bias can be measured on a few similar plots, then it is possible to correct it on other study areas counted using less thorough methods.

Sources of bias

Increased effort does not necessarily produce more accurate results. One such example is the use of mapping methods over sample surveys using transects. There has been a tendency among bird researchers to believe that results from mapping methods are close to the truth and other methods can have their accuracy compared with them. This view is now understood to be spurious. For a variety of reasons discussed in later chapters, it is often (though not always) more accurate and more efficient to use a sample survey of points or transects. Songbird surveys in the UK are now conducted using sophisticated, though simple line transect methods (the BBS) as the quality and quantity of data is higher, for lower costs.

In designing a study to handle bias it is necessary to understand its known sources. In many cases, a sensible response is then obvious.

1. The observer

Different people vary enormously in their birding skills and, more in the case of some professionals than amateurs, their motivation. No-one would put much trust in comparing counts made across several years on a reserve where the work is carried out by a new and inexperienced assistant each year. At the very least, it is essential to be familiar with the study birds, including their calls and songs if appropriate. It is also desirable to feel enthusiastic rather than run down by tiredness, cold or hunger because the schedule is too demanding.

If the count involves a number of people and they each have a turn at surveying the same plots, their differences will be spread evenly across all samples and the results are less likely to be affected. The results for each observer can then be analysed to see if any observers are less competent than others. To keep observer competence consistent, direct training may be desirable, especially if any difficult skills, such as counting large wader roosts or estimating distances, are involved. In general, we would always recommend that thought be given to training observers and ensuring that they meet a minimum standard before their results are used. If face-to-face training is not possible, then the use of full and clear written instructions can serve a similar function.

The mapping method (Chapter 3) has an additional observer influence at the map interpretation stage. Explanation of the rules by which bird territories are designated are often not explicit enough. It is therefore vital to clarify the exact methods of interpretation if several people are to do comparable analysis. Otherwise a single person could be employed to do all the analysis.

2. Census method

Different census methods vary in their susceptibility to bias, though the exact extent to which they differ, is rarely known. For this reason care should always be taken to state methods in written results, particularly if any special variations have been allowed on 'standard' methods. For instance, putting effort into finding nests can have a major effect on the results of a mapping census.

Because the method is different from the normal method, it is biased in comparison with a standard study. A more insidious way of adding bias could be the use of prior knowledge. Imagine the effects this could have in atlas studies when comparing remote areas with someone's well known local patch. Even if time spent in the field is the same, more species will be detected on the home patch because the observer knows what to expect and where to find it.

Results obtained by different methods are very likely to have different biases. In general, if there are common rules for a method, they should be adhered to. Equipment is hard to standardise but could alter results. A fieldworker with a good telescope is going to count more ducks than someone with poor binoculars.

3. Effort and speed

You generally see more the harder you try, either by walking slower or by putting in more time (Fig. 2.6). Within a study, effort should be standardised across years, plots, etc. Size of plots might equally be considered a part of effort. If plots of differing size are involved, they should receive the same effort per unit area rather than the same total effort. For common methods, there are normal standards which should be followed. As with other sources of bias, there is a trade-off between effort and results. Consider the question of what would happen to the value of the results if the effort per plot were halved so that twice as many plots could be visited. There is no magic formula for setting an appropriate effort; instead intuition and expert judgement must play their part. One way of saving on effort would be to count fewer species. Standard methods are already poor

Figure 2.6

Bias due to effort and speed

You see more birds if you put in more effort. If you race through a plot, you will miss the quiet and skulking species.

at counting some species (such as owls). So if some species are to be left out anyway, try to identify which are the critical species for study and count only those.

Effort is a particular problem in widespread surveys involving many people. For example, studies involving a large number of volunteers. If it cannot be kept constant, then the next best thing is to measure it. In this way it is possible to tell whether irregularities in distribution or numbers might be real or whether they were simply caused by variation in effort. Effort can also be published alongside results, so that a reader can judge the extent to which differences might be due to effort. In some analyses, the effects of variation in effort might be allowed for by statistical means. The UK Breeding Bird Survey, for example, adjusts results for national variation in the density of their volunteer network (Gregory *et al.* 1997). Be aware that it is just as important to record effort for plots where no birds are seen as plots that yield positive counts. If experiments are to identify the decline in the range of a species, they may rely entirely on the interpretation of zero counts. Birds of prey, for example, are often counted by checking previously known nest areas.

4. Habitat

Birds are generally easier to find in some habitats than others. If this problem is not dealt with, the methods should be designed to reduce its effect. Barn Owls are quite readily found if they nest in barns which can all be located and thoroughly checked in a study area. If they nest in trees, they will be much harder to find. Since the use of these two kinds of sites varies regionally (Shawyer 1987), there is considerable risk of bias in comparing results from different parts of Britain. A well designed study would need to recognise this bias arising from habitat differences and deal with it by using sufficient effort to find tree-nesting birds. In practice it is difficult to rule out the effects of detectability without first quantifying its effect and then adjusting the data. There are sophisticated, but easily achievable methods of dealing with the effects of detectability described in Buckland *et al.* (1993). Bibby and Buckland (1987) first showed how birds are more detectable in open habitats than scrubby ones and how this can be allowed for in estimating densities from point counts (Fig. 2.7). A review of methods to account for differences in bird detectability is found in Chapters 4 & 5.

Some habitats such as dense scrub or marshlands are difficult to count in because of sheer inaccessibility. Densities of birds in suburbia are poorly known because of a different kind of inaccessibility. Near rivers, roads or industrial sites, there may be so much noise that bird calls or songs are very difficult to detect. Even outside the breeding season, sound is more important than sight for detecting small birds in thick vegetation. It may simply not be possible to obtain accurate counts in some circumstances.

Figure 2.7

Bias due to
habitat

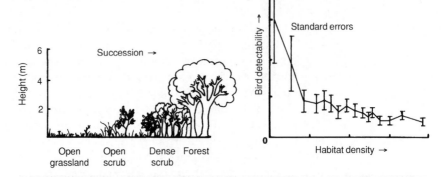

Birds are more conspicuous in open habitats than in dense woodland (from
Bibby and Buckland 1987). The hypothetical species is equally abundant
across the succession, but might appear more abundant in the grassland and
young trees where it is more easily detected. This effect is particularly serious
if the bias arises from the same source as the object of study (such as the
effect of forest succession on bird communities).

5. Bird species

Different bird species also vary in their susceptibility to being counted.
Some are noisier than others (Fig. 2.8). Some sing mainly late in the
season, some early. Some are readily mist-netted, others not. It is most
unlikely that a generalised method will count all species in similar units.
For some species, the counting method might count only parts of the population. It might therefore be valid to compare results within a species but
not necessarily between species. No general method works on all species,
and some require special methods (Chapter 8).

Figure 2.8

Bias due to
bird species

Noisy and active birds are easier to find than quiet or skulking ones. As a
result, different species may in practice be counted on different scales that do
not allow comparison with each other.

Figure 2.9

Bias due to
bird density

At high bird densities the observer may be swamped by the numbers of birds to be located, recognised and counted. It may be difficult to separate the individuals previously recorded.

6. Bird density

At high densities, the observer may be swamped by the problem of recognising different species or individuals or territory boundaries (Fig. 2.9). At low densities, boredom may tempt lack of thoroughness in searching. Counting very abundant birds poses special problems considered later (Chapter 9). Counting very scarce or dispersed species returns to the need to ensure that effort is properly distributed and recorded.

7. Bird activity

The detectability of individuals may vary according to their activity. Counting Puffins sitting at a colony would, for instance, say little about the number of birds on nests underground. In turn, the number of Puffins above ground could be related to weather, time of day or time of year. Feeding waders may be conspicuous when they are dispersed over a wide area, but when roosting, the same birds may be in a tight flock which could be overlooked. It is generally very difficult to compare counts of breeding birds in different seasons. In some cases, counting methods may be deliberately aimed at only part of the population involved in a particular activity (Chapter 8). Breeding duck numbers, for instance, are often assessed by counting the gatherings of males in the early part of the breeding season.

Colonial birds are often counted in a way that is related to breeding success. The number of birds in attendance at a colony could well be related to whether breeding success has been good or poor that year.

Numbers of Arctic Terns breeding in Shetland appear to have declined, as well they might after several years of very poor breeding success. How confident can we be that the decline is properly estimated? It is quite possible that food supplies are so poor that many individuals do not attempt to breed and, together with those that have lost eggs or chicks, they might stay at sea and avoid being counted in colonies. Seabirds are long-lived species with complex colony dynamics that include selective non-breeding and delayed sexual maturity. Pre-breeders and non-breeders may be a sizeable part of the total population but will not be seen on colony counts as they spend their non-breeding life out at sea. It has been shown that this pool of potential breeding birds will move in to fill vacant nest sites and disguise a population decline (Coulson and Porter 1987). The extent of this scenario is poorly known and may be larger than generally appreciated even in short-lived passerines. Green and Hirons (1988) provide a model which shows the effect that such a phenomenon could have on counts. In general, counters should be aware of this possible bias and consider measuring breeding success. Coulson and Porter (1987) for example, also identified a decline in the quality of breeding Kittiwakes at a colony, so that younger and lighter birds became more common. Reproductive success has been linked to age and size of some seabirds (Wooller *et al.* 1990) and the above example preceded a major population crash by several years.

8. Season

Breeding birds vary in detectability over relatively short time periods (Fig. 2.10). The best period for repeatable counts may be very brief. Many warblers, for instance, sing for a few days and then become much quieter

Figure 2.10

Bias due to season

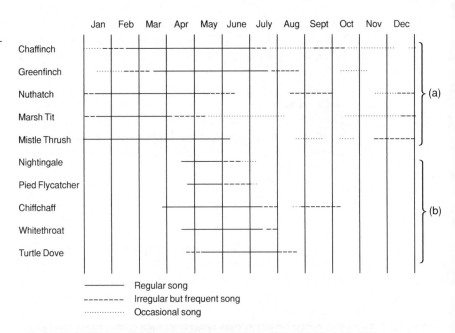

Birds sing over different periods of the year (from Alexander 1935).

(a) In the UK the resident species sing early in the spring.

(b) The resident species have all but finished singing before African migrant species arrive and begin singing.

Figure 2.11

Bias due to
time of day

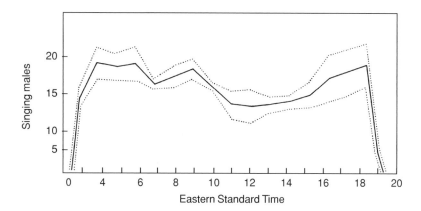

All-day activity patterns from point counts on five mid-July days. Solid line is
singing males per 20 minutes, with 95% confidence limits as dotted lines.

Activity and song output is often greatest near dawn, low during the middle of
the day, and higher close to dusk (from Robbins 1981). Places visited in the
middle of the day will therefore appear to be poorer for birds.

once mated. At the same time, growth of vegetation can rapidly make
counting harder in the early summer. Count periods therefore need to be
carefully standardised for comparison. Across years, standardisation
should ideally relate to the main breeding period of individual species,
rather than calendar days. Unfortunately, each species has a different best
period for counting, so a general census may have to compromise.

9. Time of day
Time of day should similarly be standardised because of variation of activity
(Fig. 2.11). The greatest output of song for some species is close to dawn
and is sometimes so vigorous as to be overwhelming. Again, standardisation is important. Moreover, the rate of change of singing intensity near
dawn is often high. Common advice is therefore to count songbirds starting
shortly *after* dawn. Some schools of thought do not concur with this idea
and they believe that counts should, if possible, be carried out when the
maximum number of birds can be detected and methods adjusted to
account for intense activity, e.g. by reducing the time for point counts.
Song activity will then decline significantly by some time mid-morning. The
last count should happen before this time, or both the efficiency and the
precision of counts will be seriously jeopardised.

10. Weather
Extremes of weather affect bird activity and the comfort and acuity of
observers (Fig. 2.12). High wind-speeds pose the greatest difficulties as
songbirds are harder to see and hear, especially if trees are being blown
about. Light rain, in contrast, is rarely a problem. Morning activity is often
terminated earlier in hot weather. For long-range counting of wildfowl,

Figure 2.12

Bias due to
weather

In wet or windy weather, birds may be less active and skulk out of sight. Calls
are harder to hear against the noise of the wind or rain. The observer finds it
difficult to concentrate on keeping warm and dry as well as counting birds.

waders or seabirds, light intensity and visibility are important. The best
advice is to avoid counting in poor weather. It is difficult to specify just what
this means, especially as weather factors are often related to each other and
to time of day or year. It is therefore useful to measure the weather on each
visit, preferably by categorical methods such as approximate wind speed,
direction, whether it is raining lightly or heavily, and the scale of cloud
cover from 0 to 8. This allows the researcher to make judgements about the
data and results. Allowing for it during the analytical stage is normally complicated. It is much easier to standardise counts and sidestep the problem.

Summary and points to consider

1. Precision
What is the total area under study?
What kind of study method is required?
Would a more accurate/less accurate approach with smaller/bigger
sample sizes be better?
Is it necessary to count many species or would fewer do?
How are plots/points/routes to be distributed?
Are they representative of the area being studied?`
Would a stratified design be better?
Are there enough plots/points /routes to get a sufficiently precise answer?

2. Accuracy

What steps are to be taken to deal with bias from the following?

   Observers
   Methods
   Effort and speed
   Habitat
   Bird species
   Bird density
   Bird activity
   Season
   Time of day
   Weather

Can some bias be eliminated?

Can remaining bias be spread similarly across all plots?

Can bias be measured?

Should any other factors, which might cause bias, be measured?

3

Territory Mapping Methods

Introduction

During the breeding season many birds, especially passerines, mark their territories by singing conspicuously, displaying or periodically disputing with rival neighbours.

If the density of birds is low enough, mapping the location of birds over a number of visits should result in the appearance of distinct clusters, each depicting the location of a single bird's territory. Where a species has closely packed territories, it is important to map simultaneously singing or disputing individuals. This way, if it is assumed that both birds are in their respective ranges, the boundary between their territories can be judged to fall between singing birds or at the location where disputes take place.

Territory mapping has formed the basis of the BTO's Common Bird Census (CBC) since 1962 (Williamson 1964; Marchant 1983) and has been widely used elsewhere (William 1936; Kendeigh 1944; Enemar 1959). Territory mapping is often used to derive population 'indexes'. In the case of the CBC, the population index of a species in any given start year is denoted by the arbitrary index value of 100. For instance, between the years 1966 and 1982, the population index for Stock Dove rose from 100 to 620. The results show that Stock Doves were just over six times more common in 1982 than they were in 1966.

The accuracy of population indices is influenced by bias such as weather conditions, bird detectability, and observer competence (as described in Chapter 2). Therefore, only if methods are standardised year to year or between sites, is it possible to monitor populations or compare the abundance of a species in different areas. Standardised territory mapping methods can also be used to study population densities and indeed, the method has often been seen as a standard against which other methods could be compared. This view is not always justified. Mapping does not work well on birds that range widely or do not show much territorial behaviour, especially semi-colonial species or those that do not sing, e.g. Magpie, Woodpigeon and Swift. Even among passerines there are trouble-some variations. Many migratory warblers sing for a brief time before finding a mate and then become inconspicuous. Some species, such as Linnet, nest in loose aggregations with little territoriality. Other species, such as Pied Flycatcher or Wood Warbler, may sing in more than one territory and keep quiet while moving between the two. Some species, such as Reed Warbler, occur at high densities but can move within a season if successive nests fail.

The interpretation of territory mapping data follows rules, first set out by the International Bird Census Committee (1969). When using territory mapping data, it is important to ensure that analysts are well trained and that their interpretation of the results are consistent. As discussed earlier, some species are more suited to survey using territory mapping methods, than others. For any given species, the results of a study will be biased by its biology and behaviour (i.e. whether it is semi-colonial, whether it holds distinct territories, or whether it is difficult to detect). As long as the analytical methods are well standardised (as discussed later, methods have been revised and improved to provide more consistent estimates of density) and species bias remains the same from year to year or between sites, then it is possible to identify changes in abundance. It is vital to understand that the main aim of territory mapping is to stick to these predetermined rules. This means that in any one year, the number of territories may be a poor estimate of the real population, but that does not matter, the results will still reflect changes in abundance over time as long as the analysis guidelines are adhered to.

Because mapping methods require observers to visit a large proportion of a site and conduct many visits over a season, they are an inefficient method of counting birds in terms of results per unit fieldwork effort. If it is deemed prohibitively expensive to carry out replicate fieldwork in a range of different habitats or over a large area, then point counts or transects might be considered more appropriate (Chapters 4 & 5). One advantage of the mapping method is that it can be used to depict the distribution of birds in an area. Territory mapping is often used in Environmental Assessments as it provides a convenient way of identifying areas or habitats of particular importance to species or groups of species. Spending copious amounts of time at a site also minimises the risk of overlooking scarce and/or protected species that would be threatened by development.

Field methods

1. Siting study plots
The location, number, size and shape of study plots will depend on the objective of the study and the time available. First ask yourself whether the results are to provide anything more than a simple index of species' abundance at each plot. Territory mapping can be used to illustrate spatial abundance, calculate population indexes or estimate breeding density.

If your aim is to identify the breeding population of birds inhabiting a particular area for example, then you will need a method of locating your sample plots (Chapter 1). The size and number of your plots will depend on a number of factors. If you are studying a species with very large territories, each of your sample plots should be big enough to include a single territory. If the aim is to discover exactly what species occur in an area, a large number of plots will help to ensure that the scarcest species are not overlooked. In some cases, it may even be necessary to survey the entire site although this may only be practical on relatively small areas.

Study plots should be mapped at a scale of about 1:2500 so that the positions of birds can be drawn as accurately as possible (to the nearest 10–20 m). Other scales between 1:5000 and 1:1250 may be preferred depending on the density of birds on the plot. 1:2500 scale maps may be available through local libraries. Failing this, or if the map is too cluttered, it may be worth tracing a simplified outline from a small-scale map and enlarging it using a photocopier. In open, wooded or uniform areas, it may be essential to first familiarise oneself with the plot by making preliminary visits and mapping stones, trees, trails and other features on the ground that can relate the position of birds.

In woodlands with high bird densities, a plot of about 10–20 ha is suitable for coverage in a single visit of 3–4 hours. On farmland, about 50–100 ha can be covered depending on the number of hedges and woody areas. To avoid walking on crops, plots on arable farmland will often have edges that are field boundaries. Unfortunately, this may exaggerate resulting density estimates since the bulk of the birds are in hedges. It is therefore better to have plots that are roughly square or round, and to avoid plots with long and complicated edges. Of course the effect becomes less important the larger a plot becomes, but an upper limit will be set by the time taken to cover the area properly.

2. Time and route of visits

The results of a mapping census can be influenced by the number of visits. Doing too many visits might result in some confusion. For example, if there are a lot of juvenile birds present, then it is easy to over-record a species, or confuse the interpretation of clusters. Therefore, if the survey is interested only in a particular species or group of species, it may be better to limit visits to times when the species exhibits the most territorial activity and before young fledge.

The timing of visits will depend on the location of the study. In southern English woodland, they would normally take place between about mid-March and mid-June. The CBC has adopted ten visits as a standard. Ideally they would be spread fairly uniformly at about weekly intervals. For any one species, all ten visits are rarely needed. The important visits for most resident species will fall in the first half of the survey. Later arriving migrants will not be recorded until the latter half of the season. By then, many of the residents have young, territories are vacated or not so clearly advertised and defended. The total number of visits and the length of the season simply guarantees that there are enough encounters with birds to produce clusters for all the common breeding species.

Early morning is the best time for visiting, but some (in the CBC up to two) evening visits might be helpful. It is best to avoid the first hour of activity before dawn. At this time, bird activity peaks very markedly, so there is a risk that the part of a plot covered first will produce more records. A period of more uniform activity lasts from about sunrise to about midday.

Figure 3.1

Base maps
for territory
mapping

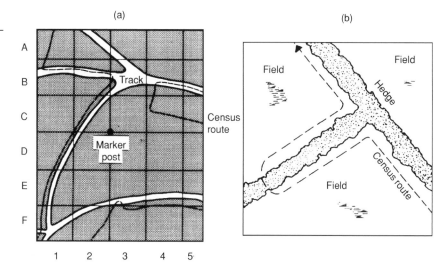

(a) Part of an outline map for a woodland plot with a 20 m grid. Grid-points are marked A1, B2, C3, C4, etc. The map also shows tracks, habitats, grids, marker posts or other features. Observers can follow the grid-lines, using a compass if necessary, so that they always know their position on the map.

(b) For studies on more open farmland, the use of the grid is unnecessary as topographical features such as telegraph poles, trees, stakes hammered into the ground at known positions, houses, etc. can be used to fix the position of the observer, and hence ensure accuracy in mapping of the registrations. In this example more birds are expected in the hedge than in the crops, and it is often difficult to walk through crops, hence the census route follows a line closer to the hedge than the centre of the field.

On hot days, these periods may be briefer. Since time of day and differences in effort can cause bias, it is important to record start and finish times as part of the documentation of a visit and to standardise these variables as far as possible.

The plot should be walked at a slow pace so that all birds detected can be identified and located. The route should approach to within 50 m of every point on the plot (Fig. 3.1). In thicker vegetation, a closer approach would be better as otherwise a significant number of birds may be missed. On farmland, all hedgerows usually need to be walked. Routes and directions should vary between visits so that there is no tendency for a particular part of the plot to be visited at a particular time. Single visits should be completed in a single period of fieldwork. Splitting visits across several days may cause problems with double recording of the same birds. If visits are split, they must be recorded as such. The duration of a visit depends on bird activity (which may be good for up to 6 hours) and on the stamina of the observer.

Figure 3.2
———————
Standard
codes for
mapping
birds

SD	Stock Dove	MT	Marsh Tit
GS	Great Spotted Woodpecker	WT	Willow Tit
S	Skylark	CT	Coal Tit
TP	Tree Pipit	BT	Blue Tit
WR	Wren	GT	Great Tit
R	Robin	NH	Nuthatch
N	Nightingale	TC	Treecreeper
SC	Stonechat	MG	Magpie
B	Blackbird	RO	Rook
ST	Song Thrush	C	Carrion Crow
SW	Sedge Warbler	RN	Raven
RW	Reed Warbler	SG	Starling
LW	Lesser Whitethroat	CH	Chaffinch
WH	Whitethroat	GR	Greenfinch
GW	Garden Warbler	GO	Goldfinch
BC	Blackcap	LI	Linnet
WO	Wood Warbler	LR	Redpoll
CC	Chiffchaff	BF	Bullfinch
GC	Goldcrest	Y	Yellowhammer
PF	Pied Flycatcher	RB	Reed Bunting
LT	Long-tailed Tit	CB	Corn Bunting

A selection of standard species codes developed by the British Trust for Ornithology for British passerine species commonly recorded on Common Bird Census plots. The full list (British Birds 1984) can be used for any kind of bird recording in all habitats and conditions throughout the Western Palaearctic.

3. Bird recording

The identity and activity of all birds are mapped with small and tidy writing in pencil or ball-point pen. If the map gets wet, some inks run; ball-points do not work on a wet map but pencils still do. It is helpful to use a standard list of codes for bird species. Codes for some British birds are shown in Fig. 3.2. Activities are also given standard codes (Fig. 3.3). Care should be taken to record as much detail as possible such as the sex and age of the bird. It is often critical in analysis to know whether a multiple observation was a party of juveniles. Were two different birds nearby a) the male and female of a pair, or b) of the same sex and thus presumably near a territory boundary?

The most useful point to concentrate on is the location of individuals of the same species that can be seen or heard simultaneously. A key feature of analysis is the assumption that territory boundaries fall between such records. For uniformly distributed species, it is difficult to analyse the results without reference to simultaneous registrations. It may not be clear if a singing bird is different from one seen earlier, a short distance away. Such ambiguous registrations must be joined by a line with a question mark to indicate that the observer is not sure whether they are the same bird or not. They are quite likely to be the same bird. If they really are different, the chances are that both will appear simultaneously on a different visit.

Figure 3.3
_____

Standard
symbols for
bird
activities

CH, CH♂, CH♀ 3 CH juvs, CH♂I♀	Chaffinch sight records, with age, sex or number of birds if appropriate. CH indicates one pair; 2CH means two pairs together.
R fam	Juvenile Robins with parents(s) in attendance.
<u>R</u>	A calling Robin.
R̳ (double underline)	A Robin repeatedly giving alarm calls or other vocalisations (not song) thought to have strong territorial significance.
ⓡ	A Robin in song.
⁃ RR ⁃	An aggressive encounter between two Robins.
＊ R	An occupied nest of Robins; do not mark unoccupied nests, which are of no territorial significance by themselves.
⊛ BT	Blue Tits nesting in a specially provided site (e.g. nestbox).
＊ PW on	Pied Wagtail nest with an adult sitting.
PW mat	Pied Wagtail carrying nest material.
PW food	Pied Wagtail carrying food.

Movements of birds can be indicated using the following conventions:

⁃ <u>GR</u> ⟶	A calling Greenfinch flying over (seen only in flight).
⒟ ⟶	A singing Dunnock perched then flying away (not seen to land).
⟶ B♂	A male Blackbird flying in and landing (first seen in flight).

(continued)

The standard BTO list of conventions is shown. These are designed for clear and unambiguous recording. Symbols can be combined where necessary. Additional activities of territorial significance, such as displaying or mating, should be noted using an appropriate clear abbreviation.

Figure 3.3
_____

(continued)

WR ——→ WR — A Wren moving between two perches. The solid line indicates it was definitely the same bird.

(WR) --- (WR) — Two Wrens in song at the same time, i.e. definitely different birds. The dotted line indicates a simultaneous registration and is of very great value in separating territories.

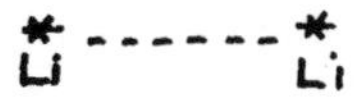

* Li ------ * Li — Two Linnet nests occupied simultaneously and thus belonging to different pairs. This is another example of the value of dotted lines. Only adjacent nests need be marked in this way.

(CK) —— (CK) — The solid line indicates that the registrations definitely refer to the same bird.

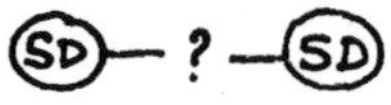

(SD)— ? —(SD) — A question-marked solid line indicates that the registrations probably relate to the same bird. This convention is of particular use when the census route returns to an area already covered – it is possible to mark new positions of (probably the same) birds recorded before, without the risk of double recording. If birds are recorded without using the question-marked solid line, overestimation of territories will result.

(WR)   WR mat — When there is no line joining the registrations, this indicates that the birds are probably different. (It is possible to use a question-marked dotted line, indicating that the registrations were almost certainly of different birds.)

C*   C* — Where adjacent nests are marked without a line, it will often be assumed that they were first and second broods, or a replacement nest following an earlier failure.

The above conventions indicate when registrations relate to different birds, and when to the same bird. Their proper use is essential for the accurate assessment of clusters.

In all cases the standard BTO codes for British birds should be used.

Some territories will overlap the plot boundary and several ways of dealing with these are considered below. Normally it is necessary to map records outside the plot. In the field, this may be accomplished by recording everything detected from the boundary without walking outside the plot. The field maps therefore need to extend by about 100 m in all directions to include these observations.

4. Habitat recording

To interpret the reasons behind the distribution of birds on a plot, it may be necessary to describe its vegetation in some way (Chapter 11). Since the bird data are mapped, it is ideal also to map the vegetation to analyse habitat utilisation by birds. If the plots are each relatively uniform and there are several of them that differ from one another, it might be sufficient to describe or measure the vegetation without mapping it.

Variations of the method

Up until now, we have concentrated on the BTO's Common Bird Census, where it is important that the method does not vary. For other surveys it may be useful to vary the method in the following ways:

1. Restricting the list of species covered

The CBC and bird community studies include all species, or at least as many as can possibly be recorded. If the purpose of the study is more restricted, the first variation is to map fewer species. This helps to concentrate time and effort in the required direction. The purpose of making ten visits is to gather enough data to register territories of species differing in their breeding and singing seasons. If only one species is being studied, five visits at the most appropriate season and time of day would suffice.

Recognising the boundaries between territories of birds of the same species is the single most important aim of territory mapping. In a study aimed at a particular species, more effort can be put into finding neighbours singing at each other or fighting. The species map could be updated after each field visit. In this way, it would be obvious if there were any areas where it was going to be difficult to resolve the number and pattern of clusters. Extra field effort could be made in these places on subsequent visits.

2. Eliciting responses

The observer can increase the chance of finding birds and obtaining simultaneous registrations by using a tape to play snatches of song of the target species and recording any responses. If a tape is played in places that may be territory boundaries, it might help to see whether or not there is a response from both birds. Owners of isolated territories often sing less than where the neighbours are close. A tape recorder could increase the chance of getting sufficient records for a cluster. Many migrants only sing early in the breeding season, but a tape recorder can get a response while birds are nesting and have largely stopped singing. Thought should be given, however, to the risk of disturbance caused by this method.

For some elusive birds, dogs can be used to increase the chance of obtaining registrations. Dogs are regularly used in counting gamebirds and for finding nests of ducks or waders. It goes without saying that a suitable and well trained dog and handler are needed if this method is to do more good for the census than harm to the birds.

3. Consecutive flush

Another way to make birds respond is to try to lure them out of their territory. This can be done by 'squeaking' and 'pish-ing' or, in some circumstances, flushing the bird out of vegetation. Obviously these methods should be used carefully. With any luck, the neighbours might be seen or heard responding at a territory boundary. These methods may be useful as a final measure to confirm a territory at the end of a survey, but should not be relied upon. Apart from aything else, it could introduce considerable bias into a survey. Much better to base initial judgements on more non-interventionist methods. Remember that once a territory has been confirmed, there is no need to further pursue a result.

4. Nest-finding

The disadvantage of looking for nests in a study involving many species is that it is unlikely that they can all be found with equal ease. In a study of a single species, the location of nests may be needed for other purposes. Since they must belong to a pair, they provide the strongest possible evidence of the presence and location of that pair.

For some species with weak signs of territorial behaviour, nest-finding is the only really good way of counting them. Some species, such as corvids or pigeons, have fairly conspicuous nests that are quite easy to find. Other nests, such as those of ducks or waders, are much harder to find. The disadvantage of using nest-finding alone is that it is extremely difficult to be confident that all nests have been found. If a territory mapping method is used simultaneously, then there is the possibility of continuing fieldwork until the results of the two methods match each other.

5. Marked birds

Much of the difficulty in interpreting maps comes from the problem of knowing which records refer to the same individual and which to different ones. This problem becomes much simpler if birds are uniquely colour-marked or radio-tagged. Territories can be drawn from mapped locations of known individuals. A drawback of this approach is that it is very time consuming. The birds have to be caught and marked, and more effort is needed in their field recording. Colour-rings can often be difficult to see and read, and radio-tracking is very time-consuming. The application of marking is further discussed in Chapter 7.

Marking is the only way of dealing with interpretation of those individuals in the population that are not breeding and are therefore behaving in a way incompatible with the assumptions of the territory mapping method. As more studies are published, it is becoming increasingly clear that many populations, even of short-lived songbirds, contain a sizeable floating population of non-breeding individuals.

6. The full study

If marking and nest-finding are used, and all efforts made to maximise records for a mapping study, there is a chance that the results will give a good picture of the absolute number of birds using an area in a season. The output will of course include much more detail than a mere count. Such studies represent the only known way to obtain absolute counts for many birds. Needless to say, they are rarely attempted because of the immense effort required.

Interpretation of results

1. Introduction
The main aim is to draw non-overlapping rings around clusters of registrations that refer to one pair of breeding birds. These clusters are often called territories and results are presented in terms of number of territories. In perfect circumstances, a species map would show a group of distinct clusters each of which would contain records of one or two birds on most visits. There would be several records of the male singing. Dotted lines would radiate to adjacent clusters to show that the observer definitely recorded different individual birds in each cluster.

Territory maps are interpreted by using the guidelines explained in the following paragraphs. It is important to be aware that the analysis does not provide a precise estimate of the exact number of birds. Differences in the behaviour of each species will result in bias, i.e. some species will be over-recorded and some under-recorded. The aim of map interpretation guidelines is to keep this bias constant over time or between sites so that changes in the index of abundance can be reliably attributed to a decline or an increase. Be aware that there may be more than one way of interpreting a territory map. If several analysts are used it is essential that they are trained to an equal standard and that they confer if there is some doubt how to best interpret an area of a map.

When maps arrive from the field, each 'visit map' will contain registrations for multiple species recorded on each visit (Fig. 3.4). The first step is to convert the data into 'species maps'. Using data from all the visits, produce a map for each species to show how the registrations form clusters. So that one knows whether two birds were recorded at the same time, each visit needs to be identified, e.g. by letters: A, B, C, D (Fig. 3.5). In practice, it is possible to plot several different species on a single sheet by the use of colours and selection of common and rare species or a pair that occupy different habitats and thus have little overlap.

Unfortunately, some species cluster well and others do not (Fuller and Marchant 1985). There will be some occasions when more than one registration in an apparent cluster comes from a single visit. The question is, is this the same bird recorded twice or a temporary intrusion of one bird into the territory of another, or does it indicate that the apparent cluster in fact refers to two pairs? There will be other areas of the map with rather sparse registrations. Are these casual records of non-breeding birds or those briefly out of their normal ranges or are they the signs of a territory where the birds were mainly overlooked for some reason? All clusters that meet a minimum standard for acceptance can be sketched in. Use a soft pencil so that changes can be made if necessary.

Figure 3.4

A field map

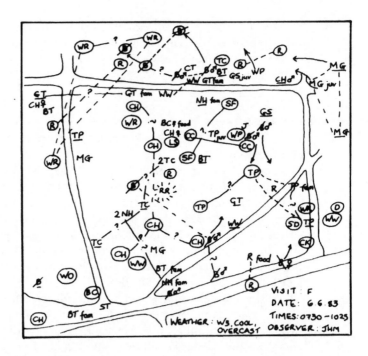

Part of a completed visit map for a woodland census, as used in the field (from Marchant 1983). This is one of the ten visits recommended for such work, spread over the spring breeding season and conducted in the early to mid morning. Such visits are labelled as A to J, and all the registrations of birds are presented in the standard way. Other important factors such as the weather, wind force (W3), date of the survey and observer name should all be appended to the survey map so that detailed comparisons between sites in the same year or the same site in different years can be made with the knowledge that other variables are not influencing the data.

In this instance it was a productive visit and all parts of the map are crowded with registrations from different species. The dotted lines will be particularly helpful in the later analysis of territories. Blackbird registrations have already been copied to the species map and cancelled with a light stroke of the pen.

What follows is a description of the standard methods of interpreting territory maps, based on rules set out by Marchant (1983) for the BTO's Common Bird Census.

2. Minimum requirements for a cluster (Fig. 3.6)

There must be at least two registrations if there were eight or fewer effective visits for the species, and at least three registrations for nine or more effective visits. The number of effective visits is the number from, and including the visit on which the species was first detected. For example, if the visits commence in late March, many migrants will not be seen at all in the first two visits. In this case they will have eight or fewer effective visits

Figure 3.5
___

A species
map

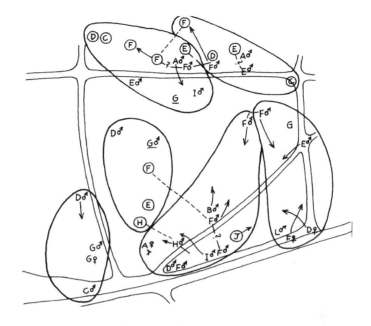

This is the Blackbird species map from the same census as in Figure 3.4. On transfer to the species map the B for Blackbird has been replaced by the visit letters A–J which represent the ten visits. However, the symbols indicating sex, song and movements have not been changed. The map has already been analysed, and six territories were found in this portion of the plot.

and will need only two registrations to define a cluster. Difficult species, especially nocturnal ones such as owls or Woodcock, are assumed to have rather few effective visits and clusters are by default, derived from just two registrations.

In addition, at least two registrations from a cluster must have been recorded at least 10 days apart. This is to avoid including temporary migrants that are present for only a few days and may be registered twice.

A single record of a nest with eggs or young can be counted as a cluster even in the event of the adults not having been seen at a level to qualify. Broods of flying juveniles or of nidifugous species such as gamebirds or waders should not be counted in the same way as a nest. They might have moved from a territory already recorded or from one outside the plot.

3. Dotted and solid lines (Figs 3.7 and 3.8)
A dotted line indicates that two different birds were seen simultaneously and should normally be used to define the boundary of two clusters (unless the birds are two adults of a pair or an adult and juvenile). A solid line is used if the observer was sure that two registrations were the same bird and should not be split into two clusters. Uncertain records joined with a question mark can be treated either way in accordance with the sense of the other information.

Figure 3.6

Minimum
requirements
of a cluster
(from
Marchant
1983)

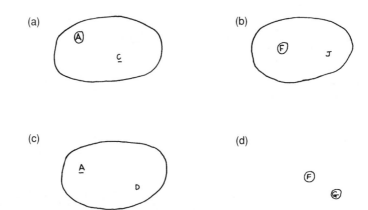

(a) There are two records of a Robin on visits A and C. If there had been only eight census visits to this plot then this amount of data would be sufficient to define a territory; however, if there were nine or the recommended ten visits to the plot then the two records would be insufficient to define a territory.

(b) There are two records of Whitethroat on visits F and J (early May and late May) out of a sequence of ten visits. Because Whitethroat is a migrant which does not arrive in Britain until relatively late in the breeding season these two registrations are sufficient to mark this as a breeding territory.

(c) There are two records of Tawny Owl on visits A and D. These are sufficient to define this as a breeding territory because Tawny Owls are difficult to count and hence the minimum requirements for a territory are lowered.

(d) There are two records of Willow Warbler on visits F and G. However, these are separated by only 2 days, not the required 10, hence the territory is not valid as the records may well refer to a bird that sang briefly whilst on passage to another site and did not breed.

It is the dotted lines that make it possible to analyse complicated maps and this is why their detection in the field is so important. Begin analysis by sketching in places that are unmistakable boundaries between clusters because they are crossed by dotted lines. Then continue to complete other clusters that contain the minimum numbers of records to be acceptable.

4. Multiple sightings
Unless there is reason to suspect that a single bird has been recorded twice, records of more than two birds (other than suspected mates or parent–juvenile combinations) should generally be treated as belonging to more than one cluster. This is where careful field observation helps. If a group of birds showed any sign of aggression to each other, then it would be reasonable to put them on a boundary between two clusters. If during fieldwork, two records fall very close together on a single visit, it is worth spending another few minutes to see whether or not they are two different

Figure 3.7

Dotted, solid and question-marked solid lines

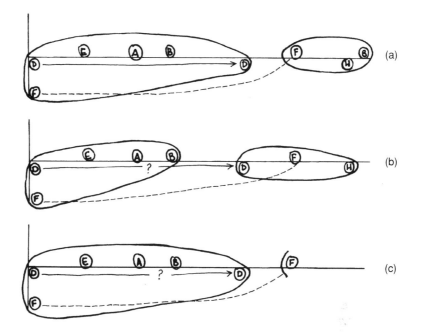

This figure shows examples of the correct treatment of lines between registrations of Willow Warblers (from Marchant 1983).

(a) The dotted line between the two warblers recorded on visit F means that the two F registrations cannot be placed in the same cluster. The solid line between the two birds recorded on visit D means that both records were of the same bird and should be placed in the same cluster.

(b) and (c) The question-marked solid line between the two birds recorded on visit D can be treated in either of these two ways, depending on the pattern of other registrations. In (b), there are sufficient registrations to support a second cluster DFH and the D records are treated as being of separate birds. In (c), there is no support for a second cluster because there are fewer than two birds recorded in the second possible cluster, and hence the D records are treated as if one bird was involved. In this case the second F registration is treated as a superfluous registration.

These examples are correct as they stand, but on a real map they might be further influenced by the pattern of adjoining registrations.

birds. This might be done by seeing whether the visible bird will fly past the place where the first was recorded, and if so whether there is a reaction by another bird. Be aware that multiple records may be migrants. For example, wintering Blackbirds are still present in Britain during April while local birds are breeding.

If a cluster contains two registrations on more than two dates and a territory boundary is suspected, consider dividing the cluster in two. Judge whether the division yields two clusters which both meet the minimum standards for acceptance and are comparable in size to other territories of

Figure 3.8

Interpreta-
tion of dotted
lines

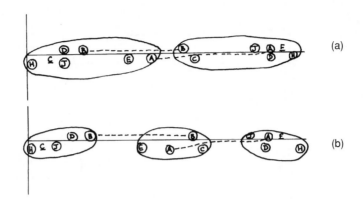

In (a) and (b) there are two different analyses of the same set of registrations
of Willow Warbler (from Marchant 1983). Standard BTO instructions to
counters state that example (a) is unsatisfactory because the apparent
nucleus of registrations on visits **ABC** is split between two clusters. Example
(b), giving three smaller clusters, is said to be a better analysis because it uses
ABC as the basis of a separate cluster. The treatment of dotted lines is correct
in both examples.

the species elsewhere on the map. Some species are particularly good at
moving rapidly and undetected across their territories. Examples are Chiffchaff, Wren and *Sylvia* warblers. Some migrants pause briefly and sing
before acquiring a territory in the spring. Extraneous records of species
such as Sedge or Willow Warbler might be due to this and do not merit a
split of a cluster. Some species, such as Blackbird and Yellowhammer move
widely outside their territories or do not have very marked territories at all.

5. Superfluous registrations
Outlying records should be included with the nearest cluster as long as this
will not make that cluster unusually large or cause it to have too many
multiple registrations. Otherwise they should be left out. Some such
records might belong to a territory largely located off the plot. In other
cases, they will belong to floating non-breeding birds, especially early in the
season or later on if they are juveniles. The analysed map should include
some notation of how such records have been treated to show that they
have been considered rather than overlooked.

6. Large or diffuse clusters (Fig. 3.9)
In uniform areas and for some difficult species, records may be widely
spread rather than grouped. It is generally best to start by drawing in the
obvious possible clusters from groups of registrations and dotted lines.
Then work from these centres to see how the rest of the data can be made
to meet the rules. This technique is much more useful than starting at one
edge of a map and working systematically across it.

Figure 3.9

Large diffuse
clusters

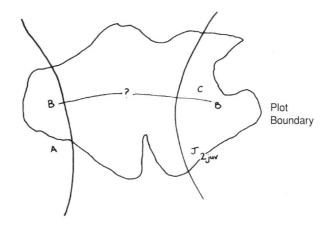

Plot
Boundary

When mapping large and mobile species such as Kestrel, it may be difficult to
gain sufficient records to be sure of territory boundaries. In this example either
one or two birds were seen on visit B. It was not known whether they were the
same bird or not. Other records give the impression of two clusters and the
interpretation of the map is the best fit of the available data. Only one territory
is confirmed.

Large species may show diffuse clusters and are more likely to have territories that reach well beyond the boundaries of the plot. In cases of large or
diffuse records, first try to locate any territory boundaries that might exist.
Then perhaps draw clusters of a size equivalent to the known territory size
of the species in question. The difficulty here is that territory size for any
one species may not be well known and may vary with population density or
habitat.

7. Spurious clusters (Fig. 3.10)

Adjacent clusters may meet the minimum rules for qualifying but should
be examined to see whether they could better be classed as one larger
cluster. The clusters should be joined as long as there have been no more
than two double registrations (otherwise, refer to number 4: multiple
sightings) of birds and the new cluster is not unusually large for the species
and habitat.

A possible explanation for two small, adjacent clusters is a shift in the
territory of a species during the course of the season. In this case, the early
registrations will tend to be in one part and the later ones in another. Such
cases should always be assumed to belong to a single pair if they still obey
the rules when joined up.

Attention should be given to the possibility that very small clusters
belong with a nearby group. Other spurious clusters without singing birds
may occur in feeding areas. If there is reason to suspect that records relate
to communal areas then they should be left out and a note made.

Figure 3.10

Territory
shifts and
multiple
registrations

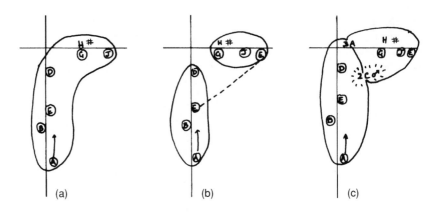

(a)   (b)   (c)

Three correct examples of analysis are shown (from Marchant 1983).

(a) The two groupings **ABDE** and **GHJ** are merged into a single cluster on the assumption that there has been a shift of territory. It would be wrong to draw clusters where such groupings are so close together.

(b) The addition of a second E and a dotted line makes it clear that there are two clusters.

(c) The figure is extended to show a correct treatment of multiple registrations. Neither cluster has any double registrations.

8. Colonial or non-territorial species (Fig. 3.11)

Because the mapping method works well only for species that show distinct clusters, there is a variation in the method to deal with semi-colonial species. Such birds may be non-territorial or defend small nesting areas but range much more widely. For these species, 'group clusters' are drawn. They include hirundines, pigeons, ducks and some finches. The CBC uses the following guidelines.

The clusters must include a potential nest-site or some other centre of breeding activity. For example, swallows feeding over a field would need to be near a building and ducks would need to be near a ditch or pond. If adjacent group clusters contain similar maximum numbers of birds on different visits they should be amalgamated as they may be the same group of birds using different areas.

Each group cluster is assigned a number of pairs based on the second highest count of males. The second highest count is used to reduce the possibility of a spuriously high temporary figure for any one visit. If the sex of the birds is not or cannot be recorded, they are assumed to be equally divided by sex, treating any 'half' counts as males.

Excessively high counts are not included in the analysis. These are any counts that may be too high owing to intrusion by temporary feeding flocks, flocks of winter visitors or passage migrants, young or post-breeding concentrations of species such as Lapwing and influxes of moulting adults, e.g. wildfowl.

Figure 3.11

Semi-colonial
species

(a)

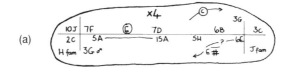

(b)

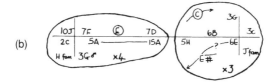

For a semi-colonial species such as Linnet it is often necessary to draw
clusters representing groups of territories. Examples (a) and (b) show correct
and incorrect treatments of the same set of registrations. Example (a) is
correct, based on totals of seven birds on visits D, E and F. The high count on
visit A is discarded as probably a remnant of a winter flock, while that on visit J
probably includes juveniles. Example (b) is incorrect, since the peak counts in
these two adjacent putative clusters occurred on different visits, and
combining them as in example (a) considerably reduced the assessment
(from Marchant 1983).

If the number of nests or broods is recorded at the same time as 'group
clusters' and this number is largest, then the nest count should be taken as
the number of pairs.

9. Edge clusters (Fig. 3.12)
If clusters overlap the plot boundary, then dividing the number of clusters
by the plot area will produce an exaggerated estimate of density. The convention recommended by the IBCC is to include edge clusters, only if more
than half their registrations lie within or on the plot boundary. Unfortunately, if the boundary is a feature such as a hedge, then all records in the
hedge will be counted as on the boundary. This method will still tend to
exaggerate densities because a species with a territory extending well
beyond the plot might still have most of its registrations on the plot
boundary. On visits when the bird was elsewhere, it would simply not have
been recorded.

There are two other ways to deal with edge clusters. Firstly, each territory
that overlaps the edge of the plot is counted as a half. Secondly, all the edge
clusters can be sketched in allowing rings of a typical size for the species in
case. The proportion of the area of each cluster that falls inside the
boundary is then included in the plot total. Proportions might be
estimated by eye to the nearest tenth for example.

These considerations emphasise the importance of dealing with edge
records properly during fieldwork. Nearly square or round plots, or plots
that do not commonly use hedges or woodlands as edges, minimise the

Figure 3.12

Edge clusters

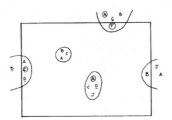

In this example of a territory map there are three possible interpretations of the number of territories depending on the system being followed. In the CBC (Marchant 1983) all the edge territories would be counted as well as those wholly within the plot; this method would therefore give five territories for the plot. In the International Bird Census Committee methodology (International Bird Census Committee 1969), edge clusters are treated differently and only those clusters in which most of the registrations occur within the plot are counted as bird-territories for the plot. In this case there would be three territories within the plot. In an alternative system only the territories wholly within the plot are counted as full territories, others are accorded a proportion. In this case the number of territories would be interpreted as c. 2.5 territories.

That the CBC regards all edge territories as a part of the total number of territories for the plot can lead to an overestimation of the density of birds on the plot unless allowance is made. Errors are worse if the margins are rich in birds. It is recommended that plots are selected so that their margins are typical of the plot in terms of their bird abundance in order to minimise this effect.

number of edge clusters with a high density of birds. The most satisfactory approach is to record well beyond the edge of the plot, perhaps by 50–100 m, which helps to interpret clusters that lie across the boundary of the plot.

Assumptions

1. The observer is good at finding and identifying birds

This almost goes without saying for any bird census method. In dense vegetation, mapping relies considerably on the detection and recognition of songs. In more open habitats, birds may flee from the observer or skulk quietly, and most of the records may have to be detected, identified and mapped at some range. The main challenge to acuity is in being able to detect and map the more distant records to maximise the numbers of simultaneous registrations. Mapping is not quite as demanding of identification skills as transects or point counts (Chapters 4 & 5) because it is possible to deviate from the route to check the identification of anything uncertain. Familiarity with the plot and its birds after a few visits also helps, compared with point counts or transects where the same areas will not be visited as often.

2. Records are plotted accurately

Inaccurate plotting greatly increases the chances that clusters will not be interpreted correctly. Plotting the location of birds is easiest if the map includes enough features that the observer can position birds relative to objects on the ground. In woodland, a compass may be essential. If the distance and thus location of a singing bird is uncertain, it might be possible to move further along the route and try again or to attempt to triangulate it with a compass. Familiarity with the geography of the plot helps. Especially in woodland, it is useful to be able to carry out reconnaissance visits to familiarise oneself with the layout of the plot.

3. The standard rules are used, or broken selectively

The factors known to cause bias such as time of day and year, weather and speed of coverage should be standardised as far as possible. The number of visits should also be close to the standard unless for good reason another number is chosen. It might be tempting to make more visits, but this generally adds more confusion rather than further useful data. The best way to maximise the reliability of records is to give close attention to simultaneous registrations during the fieldwork.

Interpretation of the maps must also be done carefully, especially when considering ambiguous or poorly defined clusters. The most subjective deviation from the guidelines is that which attempts to describe a poorly defined cluster using average territory size for the species, based on other literature or with reference to clusters elsewhere on the map. The best way out of this problem is to try to minimise the collection of ambiguous records in the first place.

4. Birds live in pairs in fixed, discrete and non-overlapping ranges

This is the most critical assumption of the mapping method and little is known about how real the assumption is. Of the birds so far studied, there is quite a range of exceptions. Many non-passerines either violate the assumption, or range so widely that mapping plots are too small to encompass more than part of a territory. Species such as doves, corvids, finches and hirundines also have patterns of ranging and territoriality not suited to the method. Polyterritorial species such as Wood Warbler or Pied Flycatcher risk being counted twice for each male. In marshes, the numbers and loose territoriality of species such as Reed and Sedge Warbler make the chances of mapping territories rather remote except in low density areas. Although a good number of songbirds do appear to breed in fixed and non-overlapping territories, there is increasing evidence that individuals of some species wander, perhaps as a result of being non-breeders. The mapping method not only fails to count these birds, it also risks obtaining spurious results because of their presence.

5. There is a reasonable chance of detecting a territory holder

Birds that sing clearly meet this assumption best, but it should be noted that the singing season for some species is very brief. Nocturnal species are also not well counted by mapping unless suitably timed visits are made. For these species, clusters with few records may be excluded because they do not clearly fit any of the criteria for defining clusters. However, these may represent real territories and the density result may be significantly low.

Examples of the use of territory mapping

1. Population monitoring in Britain

The CBC has been run by the BTO since 1962 using the mapping method. It was originated at a period of concern about the impact of habitat changes and agrochemicals on farmland birds. It aimed to measure natural fluctuations in numbers of common birds and to detect any long-term trends. Results have been summarised in Marchant *et al.* (1990). A subsidiary aim has been to generate data on bird distribution in relation to habitats.

Methods are described in Marchant (1983) and discussed in Marchant *et al.* (1990). Study plots are categorised as woodland or farmland. About 100 of each are recorded annually by observers who are mainly amateurs. Observers select their own plots and are asked to pick plots representative of farmland or woodland in the region. They are asked to follow rules which are clearly set out and are largely as set out in this chapter. The maps are analysed centrally by a small number of people who are trained for consistency.

Population indices are calculated for species with sufficiently large samples. They are based on the year-to-year changes of numbers of territories summed over all plots that were consistently covered in both years. If observers change or methods deviate from the standard, then the data are not admitted for the years involved. An index for each year is calculated by applying the percentage change from the previous year to the previous index. The index is arbitrarily set at 100 for all birds in a datum year (currently 1980). Long runs of these indices show patterns of population change over time (Fig. 3.13).

The CBC deals with bias caused by differences in observer competence by analysing paired years from the same observer. It therefore does not matter if a particular observer tends to find more or less than another would, provided this bias is consistent. There is a high level of consistency between observers in measuring year-to-year changes.

The CBC system cannot be used to represent changes in total bird numbers in the country. Nor is it known whether allowing observers to choose when they start and stop work on a particular plot, might bias the resulting population index. If plots with major habitat losses were abandoned, then the indices might not reflect the impact on overall bird numbers. The other disadvantage of the CBC is that it is very demanding of time from volunteer observers and from professional analysts. Point count and transect systems in other countries (Chapters 4 & 5) are based on more efficient methods.

Figure 3.13

Population
indices for
Song Thrush
and Blackcap
derived from
CBC data in
British
woodland and
farmland
habitats

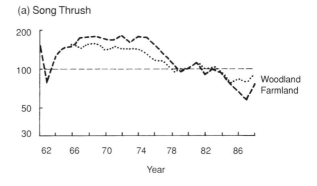

(a) Song Thrush

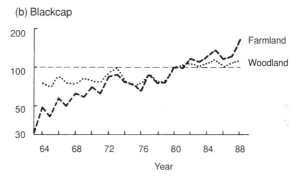

(b) Blackcap

(a) Song Thrush. This graph indicates population trends caused by both weather and other factors. The harsh winters of 1961/62 and 1962/63 caused populations of this resident British species to decline markedly. However, since the mid 1970s the population has been in longer-term decline, due in part to cold winters in 1978/79 and 1981/82 but also to unexplained factors (from Marchant et al. 1990).

(b) Blackcap. The CBC index for this migratory species shows a long-term increase throughout the period 1963–1988. There are various theories for this increase which are summarised in Marchant et al. (1990) but, as those authors conclude, no reasons can definitely be ascribed for this increase.

2. The distribution of birds in coppiced woodland in relation to vegetation age

Two studies (Fuller and Moreton 1987; Fuller et al. 1989) have related the distribution of birds to the time since coppicing in woods in Kent. A problem with coppicing is that it occurs in small blocks (medians 1.2 ha and 0.3 ha, in the two studies, respectively: range in the first case 0.3–2.7 ha) so that it is difficult to obtain samples of adequate size. The total study areas were 22.3 and 30 ha and had been surveyed for 10 and 5 years, respectively. The mapping method was used so as to associate bird records with coppice plots of known age. It would have been difficult to conduct these studies by any other method because of the small sizes of the study area.

In the first study, bird densities were estimated by attributing each

territory to the coppice plot in which it mainly lay, and summing by age
classes over all years. It is surprising, however, that it was realistic to
attribute individual territories to year classes of coppice – one would not
expect the birds to show such close correspondence to a forest plan. In the
second study, registrations were counted and summed by age classes over
years. This approach avoids the time-consuming problem of drawing and
locating territories. Sample sizes were increased by making 23–25 visits per
summer, thus deviating from the conventional use of the mapping method.
A problem with increasing the apparent sample size by making more visits
is that many records will be from the same individual birds, so one cannot
generalise from the results from the sample plots.

   Both studies reached the same general conclusions that species differed
in their preferred age classes. There was a general tendency for migrants to
be more abundant in the earlier stages and resident species at greater ages.
Results for several species are presented in Chapter 10.

Summary and points to consider

Mapped counts are time consuming to complete in the field and to analyse.
Their best feature is that, unlike other techniques, they produce a map of
distribution of birds.
Is this feature going to be used?
If not, would transects or point counts be more efficient?
Combined with colour-marking and nest-finding, mapping counts have
the potential to give good absolute estimates.
There are fixed rules for mapping censuses. Need they be used?
If fixed rules are used, they must be used strictly.
Simultaneous registrations are the key to good mapping.
Can the study be restricted to a limited set of species?
If so, there are variants of method to consider:
   Tape-recorded playback
   Consecutive flush
   Nest-finding
   Marked birds
The guidelines for analysing maps need careful consideration.

4

Line Transects

Introduction

The idea of walking about and counting all the birds detected has the appeal of simplicity. By keeping moving, it is possible to cover more ground in a fixed time than by any more elaborate method, and large sample sizes can be generated efficiently. Long transects can be divided into small sections whose habitats can be measured to assess bird/habitat relationships.

This chapter describes line transect methods used in the field for calculating relative and absolute density estimates. The method assumptions are described with hints on how to minimise their violation. Four examples are given of line transects in use. Two such studies provide widescale monitoring of birds in Finland and the UK. One example makes status assessments for seabirds offshore in the North Sea and the fourth describes the influence of habitat change on bird communities in North American shrub-steppes.

Generating absolute density estimates requires a study to be well designed, and fieldwork must follow a set of stringent rules. The main assumption of line transect surveys is that bird detectability remains constant. In reality, detectability will almost always influence density estimates between species and between habitats, e.g. if a species is particularly difficult to detect in one habitat, the data may wrongly suggest it is uncommon in that habitat compared with others. To adjust for the effect requires measurement of distances between bird and observer, 'distance sampling'. The development of the software package Distance (Thomas *et al.* 1998) (Fig. 4.6) has revolutionised biological censusing by making complex modelling of distance sampling data a practical reality, even in small projects. This analysis tool and the theory behind modelling data are described in Figs 4.5, 4.6 and 5.6 (Chapter 5).

None of the field methods for transects has been standardised beyond particular national schemes. Indeed, it would probably not be possible or desirable to standardise them because different habitats, bird species and study objectives need different methods. Unless absolute density estimates are given, lack of standardisation has the disadvantage that it can be difficult to compare results across studies.

Line transects are often used to collect data in large, open areas. This is because the method is more efficient than point counts, e.g. one tends to record more birds per unit time using line transects than point counts. In contrast, in dense habitats, it is often difficult for an observer to detect birds while moving and point counts may be preferred (see Chapter 5). However, transects are often more accurate than point counts. This is because the impact of bias rises linearly for line transects but by square for point counts. For example, transects are also less susceptible to bias caused by bird movement.

Detecting and identifying birds while walking is a challenge to ornithological skill. The approach is thus sensitive to bias from observer quality and experience. Observers' interpretation is also sensitive to variation in detectability due to bird behaviour, weather and vegetation. However, as long as birds are correctly identified and the project design and fieldwork meets the necessary assumptions, these differences can also be adjusted for using Distance software.

The theoretical basis of transect counting can also be applied to detection of signs of birds, such as droppings. It can also apply to transport methods other than foot, with specialist applications in aerial and ship-based surveys.

Field methods

1. Routes, visiting rate and travel speed

Routes are selected in accordance with the aims of the study but are usually constrained by accessibility. There is a risk that bias is introduced as a result of selecting for easy access. On farmland, for instance, it is easier to walk the field margins. This will not give a good estimate of overall densities because most species either prefer or avoid hedges and this behaviour is in turn influenced by other factors such as farming practice and timing. In monitoring exercises it is also conceivable that between year a and year b the proportion of birds feeding near field edges can change. This may be due to a decline in the suitability of hedgerows as cover habitat to escape predators or an improvement in the availability of margins as a result of conservation schemes. In such cases, real changes in the abundance of species may be disguised by other factors.

The size and position of transect routes is an important consideration. Routes should be positioned either randomly, or systematically within the study area to provide a representative sample (Fig. 4.1). Routes might be of any total length. In an ideal study, single sections would not adjoin, but all would be separate and independent (see Fig. 4. 1). In practice, spending time and energy moving from one place to another after completing a single section would rarely be practical. Instead, routes might be of a fixed length so that each could be covered in a single session of fieldwork. The total (combined) length of route also depends on the aims of the study and the resources available. Hypothetically speaking, if 60 registrations are needed for a reasonably precise estimate of density of any one species, the effort needed to estimate densities of any but the most numerous birds

Figure 4.1

Selection of
transect
routes

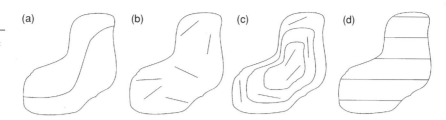

(a) A single route following a natural path is the simplest way to sample an area. The design has the advantage of being the quickest and easiest to follow in the field. It may be the only possible one if access or time is limited. It has the disadvantage that the route may not be typical of the whole area – the path may follow a feature such as a stream or a contour line which affects bird distribution and numbers.

(b) The same length of route divided into six randomly located parts. This design would give a truer representation of the bird fauna of the sampled area if that was the requirement of the study. Since the six routes are independent, it becomes possible to estimate the precision of the resulting mean counts or relative densities for species. The disadvantage of the design is that it would be harder to cover in the field and would take longer because of time spent moving between routes.

(c) Six routes are randomly located with two in each of three classes of distance from the edge of the plot in a study investigating the effects of edge and interior habitats. This stratified random design is often the best if there is some prior knowledge about factors causing variation and those factors can be geographically located in advance.

(d) Five transects are positioned systematically, an equal distance apart. Systematic designs must be used with some caution so that bias is not introduced by a parallel land/vegetation feature. It is possible to use a combination of stratified and systematic design, even incorporating a random element. For instance, transects may be set systematically within strata, but to avoid regular habitat features within strata, the start points can be randomly selected.

would be considerable. Sample size is often a limiting factor in the use of distance sampling data to achieve good density estimates for uncommon species in short-term studies.

For many analytical uses, it might be sensible to divide routes along their length into fixed intervals. Interval length would be greater in more uniform or species-poor habitats such as moorland: perhaps up to a kilometre. In richer or more varied habitats, they might be as short as 100 m. If records are separated into such divisions, some estimation of the variability of the results is possible. However, note that sections of the same continuous line are not independent of one another. One possible way of getting round this problem may be to survey using zig zag lines. Although this is not ideal, it has been proposed for boat or plane surveys where costs are too high to justify suspending a survey until one reaches the start of the next transect. If habitat features are also measured in the same intervals, then further interpretation may be possible. It is also possible to divide the transect into two halves (right and left). Counting birds on just one side of a transect, however, should be avoided as there is a marked tendency to include birds that are just on the opposite side of the line if both sides of the transect can be seen. This can cause an inflation of results at distance = 0 and errors in the estimate of density.

Routes are generally visited once or a small number of times. In a breeding survey in northern Europe, it might be sensible to have two visits to catch the peak breeding activity of sedentary and migratory species. Repeating the counts allows some chance of assessing how much more information can be gathered. In general, repeat visits will count many of the same birds and although this increases the precision of data at individual transects, because the counts are not independent, it may be better to use the time to establish more transects and improve coverage of the site.

The speed of walking the routes depends on the numbers of birds present and any difficulty in recording them all. In open habitats, a speed of about 2 km per hour might be reasonable. In thicker areas with greater difficulty in recording all the birds, half this might be reasonable. At slow speeds, it is possible that mobile birds will move into the transect and inflate their density estimate. In line transect sampling, this is not a serious concern as long as the speed of the birds is appreciably slower (less than one third the speed) of the observer. For example, to record birds in flight would require the observer to be moving at 50 km/h! Observer speed should be standardised within any one study to avoid adding bias to comparisons between years, sites or whatever.

2. Field recording
Birds can be counted within transect sections or sketched on a map. Mapping is necessary if there is some uncertainty as to how the next stage of analysis is to be completed. Mapping keeps open the option of changing some of the methods of analysis. It has the disadvantage that the data need a subsequent stage of work which could be avoided if the appropriate totals were recorded on a suitable form in the field. These days it is better to invest in a laser range finder.

There are options over what to record by way of activities. Some studies of breeding birds have attempted to get closer to estimates of pairs or singing males by recording sex and activity and treating the records differently. The potential variety of options is limitless. If there is evidence to suggest that females of a species are difficult to detect, it is best to analyse just singing males and assume an equal ratio of males to females (at least until there is sufficient data to also analyse females), or record the units as 'male territories'. The detection function for males and females combined, would be more 'spiked' and for reasons explained in Fig. 4.5, it would be more difficult accurately to model these data.

3. Distance measuring
There are many ways of generating density estimates from transect counts.

Figure 4.2

Methods for
measuring
distance in
transect
routes

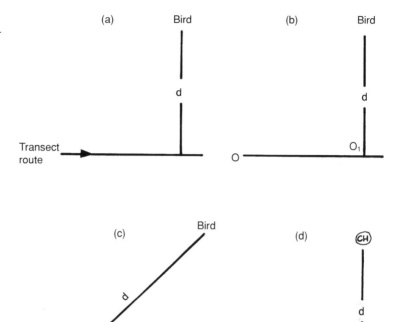

Whether distances are recorded completely or in belts, it is the perpendicular
distance between the bird and the transect route which should be measured.

(a) Distance (**d**) is estimated by eye kept in practice with periodic checking
against a measured distance. Marker posts might be set out to help.

(b) The observer (O) remembers where the bird was and measures the perpendicular distance (**d**) with either a range-finder or a tape measure when
opposite it (O_1).

(c) Distance (**d**) and angle (θ) from the route are measured with a range-finder
and a compass so that perpendicular distance can be calculated (d cosθ). This
system is not very good for birds close to the route but a long way ahead when
detected.

(d) Records are plotted on a map and distances (**d**) measured subsequently.
Good mapping is required and may be aided by fixed markers.

They all depend on some measurement of distance of the bird from the
route (see Figs 4.2 and 4.3). Distances can be estimated within belts (say
0–10 m, 10–20 m) or each measured individually. In all cases the critical
distance is perpendicular from the transect to the bird, not from the
observer to the bird. A bird that flushes 200 m ahead, but on the transect
route, counts as distance zero. A common method of recording distance is
to group distance estimates to birds into a number of intervals. The two

Figure 4.3

Different
methods of
recording
distance on
transects

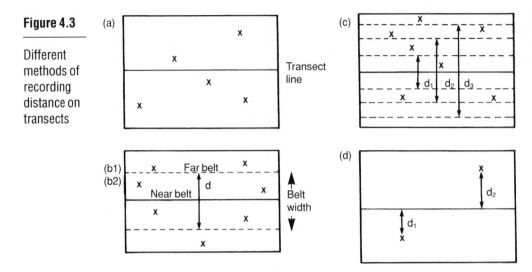

(a) No distance measuring; all birds are counted. This method is simple but
different species are counted on different scales because of differing
detectabilities. Five birds (x) have been recorded.

(b1) Fixed belt. All birds are counted within a pre-determined fixed belt (near
belt). This lowers the total count but removes distant records of the more conspicuous species. There is little to recommend this approach because it would
be optimistic to assume that differences in detectability would not still have a
large effect. If the belt was small enough to make high probability of detection
realistic for all species, then most sightings would have to be rejected. In this
case, four birds have been recorded and three birds have not been recorded.

(b2) Two belts. All birds are counted but attributed to one of two belts. This is
an effective method which is very simple to use in the field. An adjustment can
be made for declines in detectability and relative or absolute densities calculated. Four birds have been recorded in the near belt and three in the far belt.

(c) Several belts. Birds are attributed to one of several belts of fixed width
(d_1–d_3). This is harder to do in the field because distances have to be
estimated to greater precision. It is often more satisfactory to use the methods
given above or below. Counts in the first four belts were 1, 3, 2, 1.

(d) Distances are measured to all birds. Distances are perpendicular to the
route even if the bird was ahead when detected. This is the most sophisticated
method to use in the field, but generates the best data for modelling of
detection functions and estimation of densities. To assist in the accurate interpretation of distances, laser range-finders should be used where possible.

interval method is a versatile method of surveying a large number of birds
or a variety of species. It is economical to carry out fieldwork and the shape
of declines in detectability with distance (the detection function) can be
roughly estimated at the analytical stage. The more belts there are, the
harder the estimation of distance becomes in the field but the more

reliable the analysis becomes. For reasons explained later, exact distance measurement may be preferred as, since distances are usually rounded to the nearest metre, this is akin to measuring distances in very many narrow belts. By accurately collecting such data, it is possible to model the detection function precisely.

We favour estimating distances to birds in two intervals or as exact distances. Although exact distance measurement can be time consuming and costly, the availability of modern laser range-finders now makes it a simple and accurate method. If the aim is to survey a single species then exact distances are definitely recommended. In multiple species projects, it is possible to use a combination of methods and target only certain species for exact measurement. Remember that the aim of such a study is to be able to model detection functions accurately. You may find that after several days or weeks, you have enough data to do this. It then becomes less important to survey a species using this method and you may wish to begin targeting a different species.

If your data and field experience suggests that two species may have similar detection curves, then you can also apply a detection function from one species to another, or calculate a combined detection function with data from a number of species. Järvinen and Väisänen (1983a) provide a table of numbers of birds recorded within and beyond 25 m of the observer. Stone *et al.* (1995) provide similar estimates for seabirds recorded in different conditions. These figures could be used for smaller surveys to correct density estimates where the data are insufficient to estimate detectability functions. The Finnish data apply to many habitats, observers and geographical regions so there is a choice of the most appropriate corrections to use. In order to check that the Finnish results are appropriate to the new study, the near and far counts for the more abundant species could be compared with those from the Finnish samples of the same species. Since the whole philosophy behind distance sampling is to eliminate bias associated with differences in detectability, this method would be an absolute last resort and is to be strongly discouraged. Any such result should merely be a preliminary estimate, with the intention to revise it later, when more or better quality data are available. For single-species studies it is best to calculate the necessary fieldwork effort to achieve a reasonable sample before fieldwork begins. This can be done by carrying out a simple pilot study to discover the number of encounters with the species in question per unit transect distance.

4. Special variants

Various methods can be used to increase the detection rate of birds on transects. In open habitats, dogs might be used to flush sitting birds. Rope-dragging can also be used to flush close-sitting birds. This is very hard work and is only worthwhile where nests occur at fairly high densities. In some cases (Chapters 8 and 9), indirect signs such as wildfowl droppings or seabird nests or burrows might be recorded rather than the birds themselves.

As well as on foot, transects can be conducted from a car, boat or aeroplane. The same general considerations apply to such methods. Counting from a car is particularly good for large and conspicuous birds that occur at low densities, such as raptors, and in open habitats with reliably straight roads, e.g. much of Australia and the USA. However, any survey that uses artificial boundaries as transects is likely to be severely biased. This should be considered when it comes to interpreting the data.

Counting from a plane is normally used for waterfowl and other obvious or colonial species, or where access would be difficult or impossible by other means. Although expensive, aerial surveys allow quick coverage of large areas which can include the sea, inland waters and terrestrial habitats. Compared with ship surveys, planes are better to avoid double counting birds that have previously been flushed and have moved a short distance. A plane is also rather more effective at flushing birds, such as some ducks, which might be hard to see in dense vegetation. The disadvantages of aerial surveys over ship surveys are that species level identification is difficult and that it is hard to provide any additional biological or environmental information that may affect the distribution of birds. Aerial surveys are also dependent on the weather conditions being suitable both at the airport and the area of study. Air counts are usually conducted with two observers so that one can count each side of the route. Co-operation with the pilot is needed to locate records in relation to the routes. One system is to fly a pre-arranged pattern and to time the passage of landmarks, turning points and bird counts. These days it is possible to use an automated GPS system instead and end points of transects can be programmed into the plane's navigation system. A tape recorder is usually used because counts may accumulate very fast and there is no time to look down and write notes. Flights are normally conducted at a height of 50 m and a speed of about 150 km/h. Therefore, emphasis has to be placed on quick and accurate interpretation of bird numbers. For first time surveyors, identification of birds from the air is problematic and with current knowledge, many species remain very difficult to distinguish, even for experienced observers (hence some groups must be lumped together, e.g. divers and grebes). If the height is kept constant, marks on windows and struts can be used to indicate distances on the ground.

Seabirds may also be counted from a ship or boat. The highest possible forward-looking vantage point is needed. Birds that are attracted to or associate with ships, and birds in flight are recorded separately. Since flying seabirds often move faster than the boat, distance sampling methods will yield over-estimates of density so there are different methods to record them (see Fig. 4.9). As from the air, it is usual to count by time period, which can be translated to distance if the speed of the vessel is known. If the timing of each observation is recorded, the location of the ship can be derived using either the ship's or a hand held GPS system. Until recently, most ship-based surveys have not used the more sophisticated methods developed for land birds. Seabirds are generally detected because they are moving. This causes considerable difficulties as their flight speed and direction relative to that of the ship will influence the results. Gaston *et al.* (1987) provide some elaborate ways of dealing with this problem. The suggestions of Tasker *et al.* (1984) are simpler (see later). An appraisal of the advantages, disadvantages and methods of aerial and shipboard surveys in relation to waterfowl and seabirds is provided in Webb and Durinck (1992) and Briggs *et al.* (1985).

Methods of surveying using line transects

1. Fixed width strip transects

Counts may be taken either to infinity or to some pre-ordained distance. Counting to infinity has the advantage of using all possible bird records. The disadvantages may be that more distant birds were not in the same habitat as those recorded along the route. Different species are counted on very different scales by such an approach (Fig. 4.4). Whether or not this is a sensible design will depend on the objectives of the study and the nature of the ground/sea covered.

Counts within a fixed width or strip transect give smaller numbers but they have the advantage (if this is needed) that the birds are all within the habitat described. Although the effect may be less than with counts to infinity, different species will again be counted on different scales. Although it is sometimes done, there is rarely justification for dividing the count by the area (length of route times strip transect width) and calling it a density. This assumes that all birds are detected and none has fled from or been attracted to the observer. Such an assumption might be valid if the

Figure 4.4

Differences in detectability between species

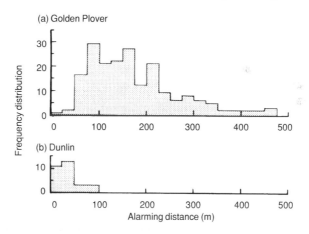

(a) Golden Plover

(b) Dunlin

Frequency distribution

Alarming distance (m)

Numbers of (a) Golden Plover and (b) Dunlin, detected at different distances from the observer on moorlands in northern England are shown (from Yalden and Yalden 1989).

Golden Plovers are noisy and conspicuous, especially as they react to an observer and give alarm at greater distances than do Dunlin.

Dunlin are cryptic and sit tight, so are not detected beyond 100 m. This difference between species needs to be taken into consideration in the design of appropriate breeding water sampling methods.

strip is narrow and the counting unit is nests of a colonially breeding species (Chapter 9), but even in such a case, detectability falls off very fast. Thirteen per cent of duck nests were missed when an observer searched a strip of 2.46 m on either side of a transect. The strip would have to be 1.54 m on either side for 100% efficiency (Burnham and Anderson 1984). For live birds of most species, an assumption of complete detectability at any distance from the observer could not generally be warranted.

2. Distances grouped into two intervals

If distances to birds are grouped into two intervals, it is possible to assume the general shape of the relationship between distance and detectability and use this to estimate density. Of the simple models available for such data, the most plausible is the half normal function. In this, detectability falls off slowly at first, rapidly at some distance and then again more slowly. The half-normal function is commonly applied to the two belt transect method as in the UK national BTO/RSPB/JNCC Breeding Bird Survey (Gregory *et al.* 1997). Theoretically, this method can be used to derive absolute density estimates. However, there will be some bias associated with the generalisation of the model for all species, so it may be better to assume estimates are relative.

Distance software yields standard errors of density estimates from two-belt surveys. However, a short cut is given in another useful paper (Järvinen and Väisänen 1983b) which shows that the standard error can also be predicted remarkably well from the estimated density, the number of transect routes involved and the correction factor (as above). The latter has only a small effect. The standard deviation is approximately proportional to the square root of density and inversely proportional to the number of routes counted. The effect of the correction factor is again small. Järvinen and Väisänen showed that their conclusion was correct in an independent test far to the north in Finland and in different circumstances from the source of the original data. Their equations could thus be used with reasonable confidence elsewhere in Europe. They can also be used in planning future work.

3. Full distance measuring

If distances to all birds are measured, it is possible to make a more detailed model of the shape of the plot of detectability against distance from the observer and reliable estimation of absolute density may be possible. The software package Distance (Fig. 4.6) has made the practice of modelling biological density data accessible to most researchers. Exact distance estimates have a marked tendency to be rounded to end in a zero or a 5. This may be avoided by deciding in advance to estimate distances to the nearest 5 m. This may not, however, be a reliable method. It is important to measure near distances precisely as small deviations cause relatively high variation. If detectability declines sufficiently quickly, it may also result in a heaping of observations at zero and an exaggerated density estimate. Heaping of observations caused by rounding figures to the nearest 5 or 10 m can affect the fitness of model estimation in Distance software but can

Figure 4.5

Analysis of
line transect
data using
Distance 3.5
software

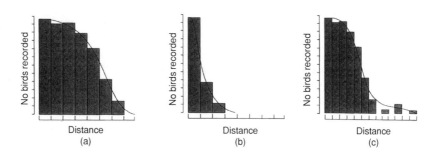

Line transect and point count sampling methods have been described in the book **Distance Sampling** (Buckland et al. 1993) along with statistics that form the framework for the accompanying analysis software, Distance.

Distance can be used to apply four basic models (key functions): uniform, half-normal, hazard rate and exponential. The choice of key function may depend on a visual inspection of the shape of the data after outliers are truncated (see below). If necessary, the function can then be fine-tuned to improve its fit using series expansions: cosine, simple polynomial or Hermite polynomial. To choose which model to use, Distance provides ways of testing model fit, of which Akaike's Information Criterion (AIC) is the preferred method (see Buckland et al. 1993). As long as the model fits a 'shape criterion' (see below), the key-function/series expansion grouping with the lowest AIC is the model with the best fit. By testing the fit of a variety of models, Distance does not assume that the function of decline in detectability with distance is the same for each species. Assuming that the method assumptions are met and data are collected properly, this allows the researcher to calculate reliable absolute density estimates. Another useful test is the Chi-squared statistic. A significant goodness of fit statistic warns that the model may be poor or that one of the assumptions might be seriously violated.

Other variables that confound differences in density estimates, e.g. habitat type, observer, weather conditions or species characteristics such as age and sex can also be differentially adjusted to analyse their effects. Generally, if $g(0)$ (the probability that an object on the line or point is detected) = 1, then you only need to account for differences in detectability if you want to estimate density within variables, e.g. if you want to estimate density in individual habitats then you should use Distance to check whether the detection function differs in each habitat stratum (by stratification or post-stratification). If, on the other hand, you just want a density estimate for the whole site then Distance draws a pooled model which should still yield a reliable estimate of overall density.

As long as a study is well designed and meets the necessary assumptions, Distance can be applied to most types of biological data to derive relative or absolute density estimates. The program also includes features to deal with 'clustered' observations (see Fig. 5.6), and incorporates a 'bootstrapping' facility to derive more robust variance estimates and confidence limits.

Distance provides an opportunity to improve the model fit by organising the data to meet the following criteria. Do not, however, rely on data manipulation to salvage a project that was not designed or implemented properly in the first

Figure 4.5

(continued)

place! If modelling theory is to provide a robust estimate of density, the curve must meet a 'shape criterion'. Broadly speaking, the detection curve must have a 'shoulder' – this is one reason why the exponential model is not a reliable key function, even if it minimises the AIC value. Detectability must be certain, or near certain for some distance. Heaping of observations around zero distance, causing a spiked distribution, can result from a number of project design flaws, but may also be a true characteristic of the data (b). If the latter case proves to be true, the fieldwork method could be changed. For example, if females are shy and elusive and both males and females are recorded, then the curve may be spiked (because a large proportion of females are only recorded near to the observer). It may be better to model data from singing males and either assume the number of males is equal to the number of females, or record the units as 'male territories'. Or, if collecting exact distance estimates, the cut off points of distance intervals could be changed so that there are at least two intervals on the 'shoulder' of the curve.

Secondly, outlying observations beyond the normal range of data should be excluded. For example, a bird seen singing above woodland canopy on the far side of a valley will not provide useful data for modelling that species detection function. These 'outliers' adversely affect the fit of models and the resulting density estimate. Truncation is standard practice for almost all distance data sets (Buckland et al. 1993). Often one truncates the furthest 10–15% of the data, but it will depend specifically on the spread of distance data for each

Figure 4.6

Using
Distance 3.5

Distance evolved to meet requirements for software to analyse distance sampling data from line transects, point transects and cue counts. Anyone attempting to carry out a distance sampling project is strongly recommended to use Distance. Not only is it fairly simple to use, but it offers the opportunity to carry out complex analyses quickly and is fast becoming the standard among biological surveyors.

GI Distance 3.5 (Thomas et al. 1998) is the result of a project to update old software in answer to the demand for a 'point-and-click' Windows based program that is more user friendly than the previous command-based versions. Distance 3.5 does not offer any major new analysis opportunities (this is an aim for Distance 4, scheduled for release in Autumn 2000) but there are a few minor changes to the way that data are processed and the results are more accessible and presented more clearly.

One of the biggest advantages both to new and old users of Distance is the facility to import data from external sources. In addition, the detailed results output is split into pages and includes a high resolution plot of the detection function. Finally, the program includes Help files that are context sensitive and provide a comprehensive guide for users.

The Distance Sampling Handbook (Buckland et al. 1993), Distance 3.5 and earlier DOS versions of Distance are available freely to download via the Distance home page. The web-site address is:

http://www.ruwpa.st-and.ac.uk/distance/

Figure 4.6
_____

(continued)

To keep up-to-date with information on Distance, it is recommended to check the Distance home page and subscribe to the e-mail discussion group. This is done by e-mailing mailbase@mailbase.ac.uk and typing **join distance-sampling** <your first name> <your surname>.

Setting up a new file in Distance

The following is a quick introductory guide to using Distance 3.5. Where we have had to refer to an example, we have used a simple case where clusters of birds were recorded in two distance bands of 0–25 m and 25–100 m.

On opening a new file in Distance, you need answer the following questions by ticking the appropriate boxes:

• Was the survey Line transect/Point transect/Cue count?

• Were data collected as Perpendicular distance/Radial distance and angle?

• Were observations Single/Clusters?

• What is the sampling fraction? – Typically this will be 1.0. If for example only one side of the transect was surveyed, e.g. during seabird counts from a boat, then the sampling fraction would instead be 0.5.

It is then possible to set the units of original measurement and the units for presentation of the results.

The main menu

Once opened, the main menu is displayed from where you can access the two main utilities (Data Explorer & Analysis Browser) by clicking on the appropriate icon:

The Data Explorer

The Data Explorer is used to enter or manipulate data. To use this facility, it is first necessary to understand the four data layers used in the Distance database.

Global layer: This layer contains data on the whole of the study, including multipliers (e.g. cue rate if the project is a cue count project). When importing data, this defaults as the name of the data set file but the label can be changed by clicking on the Study Area data layer.

Stratum layer: This layer is used to divide the data into one set of strata, for example, habitat or geographical location. Column 'Area' is used to define the size of each stratum. This is essential if Distance is to calculate abundance or a pooled estimate of density.

Sample layer: A label for each transect, point or cue is illustrated, along with the effort at each sample site. In the example shown, five visits were made to each 1 km transect.

Figure 4.6

―――――――

(continued)

Effort (or the effective length of each transect) is set at 5km so that Distance knows to divide the final result by five.

Observation layer: In the example below, data were collected in two distance bands and as clusters of birds. It is therefore necessary to arrange the data in two columns: perpendicular distance from the line and cluster size. For convenience sake, perpendicular distance is recorded as the midpoint of each band (0–25 m and 25–100 m), although these figures could fall anywhere within their respective categories as we chose to set the intervals manually (see below). The third column is species. Each time a separate analysis is created, it is possible to filter the database by this column.

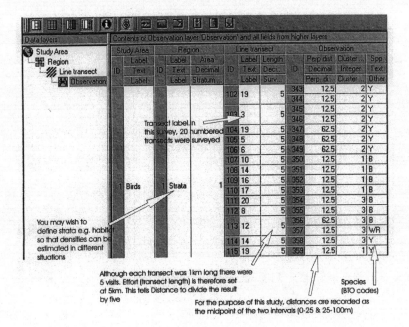

You can view your data in the Data Explorer. In this window, there are options to lock the dataset so that no changes can be made to the spreadsheet.

The Data Import Wizard

Data can be input by hand, straight into Distance, or imported from another program. Data can be imported as an ASCII file (columns separated by tab, comma, semi-colon or space) using the data import Wizard. The procedure is straightforward enough as long as you make sure that you have columns for sample label (transect, point, etc.) and distance. The following is an example of the data we arranged for import into Distance:

Figure 4.6

(continued)

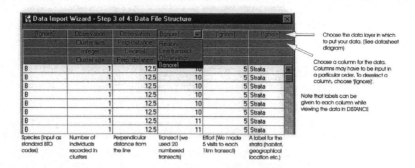

When importing data, first choose the layer and then the field for your data. Once you have assigned data to a field in Distance, its name disappears from the drop-down menu. You must assign data to each of the fields listed in the drop-down menu and the program will insist you do this before importation of data can proceed. If your ASCII file does not contain such data, then go back to your original spreadsheet and create it. If you assign data to a field by accident, choose '[Ignore]' to reset the entry boxes.

The Analysis Browser

From this window it is possible to create and run analyses, arrange, sort and view the summaries of results.

To create a new analysis, click on the icon. This opens the Analysis Details window (above) where it is possible to set up and run analyses, view the results and the underlying command structure of Distance.

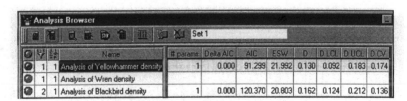

Analysis details window (inputs tab/results tab)

Once you have run an analysis, the full results are displayed by clicking the 'Results' tab of the Analysis Details Window and can be searched using the drop-down menu.

The figures that follow show examples of the Inputs and Results windows from the Analysis Details Window. These options are toggled between using the tabs on the right hand side. You can establish a number of data filter and model selection frameworks to choose from each time you conduct an analysis. Choose 'New' from the data filter display and you are given options to:

Figure 4.6

(continued)

Filter your data by any one of the fields in the Data Explorer.

Set intervals for the data analysis (in our example we chose manual intervals at 0, 25 and 100 m).

Choose how the data are to be truncated.

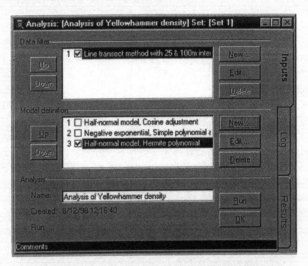

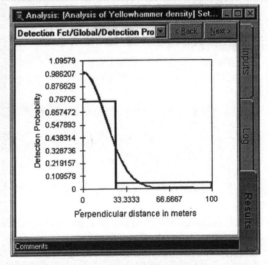

Under model selection there are options to choose for:

- Level of data output (e.g. any combination of encounter rate, density, cluster size and detection function)
- Stratification
- Model type
- Treatment of clustered observations
- Other statistical measures e.g. variance.

be avoided using laser range finders for distances over about 20 m. Best practice is to measure exact distances and if any grouping should occur, then it can be done during the analytical stage. A description of these methods is given in Fig. 4.5.

Assumptions

Setting an adequate sample size will depend heavily on spatial variation in bird density. Power calculations can be used to estimate the appropriate sample size to achieve certain levels of precision. This relies on encounter rate data from a pilot or previous study and a measure of data variation. Formulae for calculating appropriate transect line length or the number of points are described in Buckland *et al.* (1993). As a guide, a minimum sample size for satisfactory description of detection curves may be 100 detections, though it is possible to model data with lower sample sizes. For many less common species in a community, this requires a very much higher effort than will generally be given in a study. This problem can be overcome by using data from elsewhere for the species or from other species believed to be similar with respect to detectability.

1. Birds exactly on the route are all detected
The most sophisticated of methods of absolute density estimation do not work if birds directly on the transect line can be overlooked. In reality, to be sure that models are appropriately chosen and subsequent density estimates are accurate, all birds on or near the transect line must be detected (see Fig. 4.5). This might happen if birds are very cryptic in trees above the observer or because they have fled unseen at some distance ahead (thus violating the next assumption as well).

2. Birds do not move before detection
All variants of density estimation assume that birds are randomly distributed with respect to the distance from the route, i.e. non-random distribution by itself is not a problem, as long as transects are placed at random. If the birds move in response to the observer, then this will cease to be true. Songbirds are commonly attracted to observers, but in some situations and for some species, fleeing may be a more likely reaction. In these cases, it is often found that rather more birds occur at some distance from the observer than close to the route. To meet both the first and second assumption, particular care should be given to the area immediately ahead – taking care not to under-record birds away from the line – from which birds are most likely to flee when they detect an approaching human. In methods using two or a few belts, it is only important that birds do not move from one to another, so the assumption is slightly less restrictive.

Assumptions about movement might also be violated if birds move a lot in general and are only detected when they are quite close to the census route. If the route was walked at an ever slower pace, more and more such birds would be detected merely because there was a higher chance to do so. Observer speed should, however, remain relatively quick compared with birds if accurate absolute densities are the aim of the study.

3. Distances are measured accurately

Accurate distance estimation is the foundation of line transect sampling, its importance cannot be overstated. However, it is not easy to do in the field. For birds detected well ahead, it may sometimes be sensible to note a nearby feature and estimate the perpendicular distance from the route once the point is reached. A tape measure can be used, but this is very time-consuming. An optical range finder is more expensive but very helpful up to about 100 m. In projects where birds are commonly recorded over 100 m survey lasers are useful to determine distances accurately. At about $300, their cost is soon justified by the ability precisely to determine distances up to several hundred metres. Combined with a good quality compass, angles can be estimated to calculate accurate perpendicular distances. If distances are visually estimated, then thorough training and repeated self checking against measured distances should be used. One or two weeks of training is generally recommended before beginning a survey. For belt methods, the bird only has to be recorded in the correct belt so the skill demanded in the field is that much less daunting.

4. Individual birds are counted only once from the same line

On its own, repeat counting is theoretically acceptable. For example, counting a bird twice from two independent transects when transect width = ∞ does not violate any of the method assumptions that are described later. A deviation from the assumption that birds do not move in response to the approach of an observer may be caused, however, if a bird is flushed from one transect onto the next. If a species is very abundant there is a risk of double counting from the same line because of sheer confusion of birds appearing in all directions. Species that move quietly between places where they sing or call might also cause such confusion. The only advice that can be given for fieldwork is to try to keep track of individuals of the more difficult species. Avoiding problems from this assumption also calls for moving faster rather than slower down the route but this is usually constrained by the need to meet the previous assumptions, which demand more time.

5. Individual birds are detected independently

The main difficulty with this assumption is likely to arise if birds are more detectable at high rather than low densities. One bird singing or giving an alarm call might, for instance, stimulate others to do so. At lower densities, this is less likely to happen so individuals may be less likely to reveal themselves. This effect may be resolved by stratifying the data by areas of high or low density, e.g. including covariates in the analysis. Although the analysis of this method is currently too sophisticated for most researchers, it will be possible when the next version of Distance is completed.

6. Observer and seasonal effects

Transects are more dependent on observer skill than mapping counts where repeat visits give a better chance of finding each pair of birds, and a territory can be registered even if the occupant is overlooked on several visits. If there is reason to suspect that observer competence may be introducing bias into the results then it is useful to adjust this difference using Distance to stratify the data by observer. Seasonal effects might also be large when comparing across years, especially if counting sometimes extends into periods when juvenile birds are unwittingly recorded. Routes should ideally be revisited at the same season. Ideally this would be phenologically the same, but in practice calendar date might have to be used if, for instance, timing of breeding is not known.

Examples of the use of line transects

1. Population monitoring in winter in Finland

Population levels of both wintering and breeding birds have been monitored in Finland by transect methods. The winter project started in 1956 and is the only long-running study of its kind in Europe. Methods and results are described by Hildèn (1986, 1987). The aims have been both to monitor population trends and to describe winter numbers and distribution in relation to habitat.

The field methods are very simple. Observers choose their own routes with general guidance. Three counts are conducted each winter within defined 2-week periods. All birds are counted irrespective of the range at which they are detected. About 600 observers take part each year and routes average 11 km in length. Over 10 000 individuals are counted annually for the most abundant species. The maximum annual count of the scarce Black Woodpecker is 91. The simplicity of the method is believed to contribute to its popularity and thus to the amount of data gathered. Results are expressed as individuals per kilometre walked. They can be divided regionally.

Although confidence intervals are not given for individual estimates of this density index, they could be calculated from the data. Inspection of results (Fig. 4.8) shows that year-to-year changes are often quite small for non-irruptive species such as the Willow Tit. This suggests that the relative densities have been measured quite precisely. Other species show long-term trends or irregular patterns associated with variation in food abundance.

This is a simple study design with methods adequate for its purpose. Open winter habitats and low bird densities lend themselves to transect counts. Because no distance measures are used, the counts cannot be compared across species. Comparisons across habitats would also probably be biased because a higher proportion of birds present would be recorded in more open areas. For population monitoring, these problems do not matter so much. There could be a problem, however, due to observers choosing their own routes. Population changes are not computed on a year-to-year basis for the same routes as in the British Common Bird Census (see Chapter 3). If habitat changes were severe, it would be possible that population trends could be biased or for their effects to be concealed by observers picking more interesting areas to count.

More recently, Finland has introduced another wildlife monitoring scheme (see Fig. 4.7). The wildlife triangle scheme (Lindén *et al.* 1996) is a standardised survey based on three, 4-km transects arranged as an equilateral triangle. The advantage of this arrangement is that it avoids non-random bias introduced by the common northwest–southeast arrangement of glacial valleys. The scheme is aimed at hunting reserves, and with 3000 clubs in Finland, they hope to have surveyed 1200 triangles nationwide by the year 2000. The project differs from earlier censusing in that it assesses wildlife populations in relation to habitat.

2. BTO/RSPB/JNCC National Breeding Bird Survey (BBS)

Heavy declines in the populations of many common species in the UK (Fuller *et al.* 1995) have influenced the development of the national BBS, an extensive scheme to monitor bird populations (see Fig. 4.7). The project has been running since 1994 and is based on standardised survey of 1-km squares randomly selected from the national Ordnance Survey grid and stratified by observer density (Gregory *et al.* 1997). Like the Finnish example, the BBS relies on a massive network of volunteer ornithologists. Although it is more complicated than the Finnish scheme, the effort involved is still substantially less than alternative mapping techniques such as Common Bird Census (CBC).

BBS is an efficient way of recording data, and fieldwork is simple to carry out. Two visits are made to each square to record birds during the breeding season. These comprise an early (April–mid-May) and late (mid-May–June) visit to maximise the chances of recording early breeding resident and late breeding migrant species. Birds are recorded each side of two parallel 1-km transects that are each divided into five 200 m sections. All contacts to birds are recorded and placed into one of three distance categories: 0–25 m, 25–100 m, 100+ m or as in flight. Transects are walked soon after dawn, avoiding the initial period of intense song activity. During a preliminary visit, habitat characteristics are described for each of these 200 m sections.

By recording data in distance intervals and describing habitats for each transect section, there are diverse analytical opportunities in BBS. The national scheme uses Distance software to remove the effects of different detectability between species on density estimates. It would be unnecessarily complicated to interpret data recorded to an infinite distance, so for the purpose of density estimation, observations beyond 100 m are ignored. Although Distance is capable of applying a number of simple models, a half-normal detection function is preferred as it is commonly applied to binomial data of this sort.

Population changes can be assessed by comparing density estimates over time. One concern is bias introduced as a result of walking field edges. This may result in poor estimates of density due to spurious modelling of

Figure 4.7

Examples of
two national
bird monitor-
ing schemes

(a)

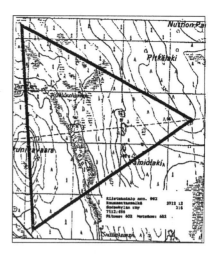

The Wildlife Triangle Network in Finland (adapted from Lindén et al. 1996)

(b)

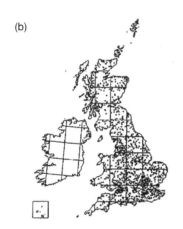

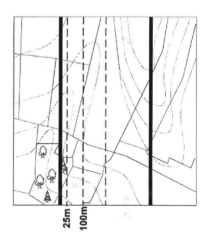

25m
100m

The distribution of BBS squares in 1995/6 (adapted from Gregory et al. 1997)

(a) The Finnish Wildlife Triangle Scheme is a standardised monitoring project
that uses three, 4 km transects arranged as an equilateral triangle. The survey
is aimed at the 3000 hunting clubs in Finland who provide skilled identification
experts to voluntarily count birds and animals. The triangular arrangement of
transects avoids the problem of bias associated with the regular
northwest–southeast arrangement of glacial valleys in Finland.

(b) The UK Breeding Bird Survey (BBS) randomly selects 1 km squares and
uses volunteers to conduct annual, standardised surveys of birds on two
parallel 1 km transects at least 250 m apart. BBS is a distance sampling
method where birds are recorded in two bands, 0–25 m and 25–100 m, on
both sides of the transect.

In both cases, estimates of national population would need to be weighted for
effort in different parts of the country.

Figure 4.8

Population
trends of
Willow Tit and
Capercaillie in
Finland in
winter
measured
from transect
counts (from
Hildén 1987)

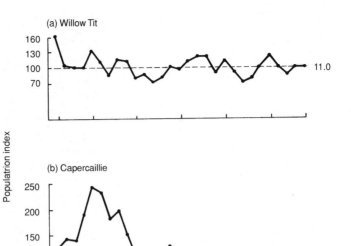

(a) The population of the Willow Tit has been stable since the mid 1950s.

(b) The population of the Capercaillie has been in decline since the early 1960s.

detection functions. In addition, with about 70% of land in the UK in agricultural production, change in the suitability of field edge habitats to birds could be a major factor producing biased estimates of decline or increase in abundance over time.

3. Distribution of birds in the North Sea

Assessing the distribution and abundance of seabirds at sea has been much in demand especially in relation to recognising places and times when birds are at risk from oil spillages. Birds may be counted by transects from ships. This is not easy because some species are attracted to ships and others flee. Flying birds violate a critical assumption. If they were all counted fully, a stream of birds crossing the ship's route would lead to an inflated estimate of density. Small or dark species are also much harder to detect than the larger and paler species. Those, such as auks, that sit on the sea are harder to detect than those, such as shearwaters, that habitually fly so it is important to model detectability for these species.

There are now standard methods which have been widely used around British coastal waters (Fig. 4.9) (Tasker *et al.* 1984; Webb and Durinck 1992). The first method comprises counts of birds that are on or 'using' the sea, e.g. dip-feeding petrels, during a standard time period in a 300 m wide distance band. This is subdivided into four discrete belts (0–50 m; 50–100 m; 100–200 m and 200–300 m) plus birds recorded beyond 300 m. Over time

Figure 4.9

A counting
system for
seabirds at
sea (adapted
from Webb &
Durinck 1992)

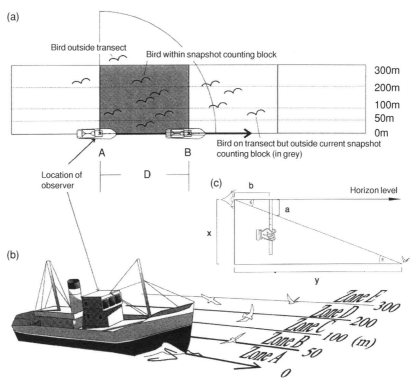

(a) Seabirds are counted in two ways. 1. Distances to seabirds are estimated in several zones so that detection functions can be calculated for each species (in this case, 0–50 m, 50–100 m, 100–200m, 200–300 m, 300 m+). 2. 'Snapshots' of the number of birds flying in rectangles of known area are made. For example, when the observer arrives at position A, he/she estimates the number of each species in the shaded rectangle. The distance D is the distance to which the observer can reliably detect all the birds. Assuming the boat is going at a constant speed, the time to reach point B from point A can be calculated. When the boat reaches point B, a count of birds in the next rectangle is made. This way, it is possible to estimate the density of flying birds.

(b) For distance estimation, the observer sits at the top of the boat and estimates the horizontal distance to all birds. Normally seabirds are counted over a 90° angle, i.e. only on one side of the boat.

(c) Estimating horizontal distances is made easier using a calibrated ruler or a pair of calipers. The top is placed level with the horizon and the distance (a) is measured. Distance (b) is already known as the ruler or calipers are strung at a fixed distance from the neck of the observer with a piece of string, which is kept taut when the readings are being made. Distance (x) is the height of the observer's eye above the water. To calculate the distance (y) precludes the use of trigonometry since the ratio of a:b is simply equal to the ratio of x:y. Therefore $y = bx/a$. There are more complicated methods which take into account the earth's curvature for instance, but for the purposes of estimating distances to birds up to 300 m from the boat, this simple method will suffice.

it is then possible to model detection functions for each species. For example, Stone *et al.* (1995) have published adjustment factors for several seabird species in different sea states. To maximise the sample of data collected, birds should be detected with the aid of binoculars. Often, two observers are used, one to 'guard' the centre line (and make sure that no birds on the line go undetected before flying off) and one to record birds in the outer bands. Counts are conducted every 10 minutes, although this could be less in areas where bird density was particularly high. The distance covered is calculated from the average speed of the vessel and densities represented as birds per unit area.

The 'snapshot' method is carried out simultaneously and, since the majority of species are recorded in flight, aims to record the density of flying seabirds. During each snapshot, the number of flying birds is recorded in an imaginary box 300 m wide and as far ahead as the observer deems suitable for the species and conditions. The time between these counts, i.e. the time it takes to travel to the start of the next snapshot, is determined by a simple formula relying on this preconceived distance and the speed of the vessel:

Number of snapshot counts per 10 minute interval $= (V \times 0.309)/D$
Where V is the ship's speed in knots and D is the maximum detection distance for flying birds in kilometres

If the birds are flying in one general direction, this method can be modified. Counts are made for a minute in a similar imaginary box and the time for an individual to cross it is also estimated. Such counts can then be turned into birds per unit area. Densities of flying and sitting birds can be added. Birds associated with the ship may be recorded but are not added to the estimates of densities.

This is the only method to date that is widely used to estimate densities of flying birds at sea. Unfortunately, it precludes modelling detectability and is therefore subject to serious bias and cannot be used to obtain absolute density estimates. To do better would be very much harder and would require more standardisation of viewing conditions than is realistically possible. The essential feature of design is that methods have been standardised as far as is practicable. If everyone followed such a standard, counts of birds at sea would be more widely comparable than is presently the case.

A similar method used in Canada is described in Diamond *et al.* (1986).

4. Habitat of shrub-steppe birds in the USA

Bird communities of the shrub-steppes have been studied using transect methods (Rotenberry and Wiens 1980; Wiens 1985). Individual routes were walked four times in mid-June when bird detectability is at its highest. Records were grouped into several intervals up to a range of 244 m on either side of the transect. Habitat data were recorded by laying ten perpendicular transects at fixed intervals across each route and recording detailed data at one randomly located quadrat within each 10-m length of the cross transects. The main studies relied on having various different plots surveyed in the same way. The fairly open habitats lent themselves to transect counts. Mapping surveys would have been far more time consuming.

In one study (Wiens and Rotenberry 1985), the effects of a herbicide treatment and reseeding were reported. The affected plot was surveyed for 3 years before and 3 years after the treatment. Other plots in the area acted as controls. This was not a designed experiment; the authors did not know that their plot was due to be 'improved' but took advantage of the event.

Changes of abundance of four species are shown in Fig. 4.10. Two analytical approaches were tried. The authors predicted the effects to be expected knowing the vegetation changes that had taken place and having studied the relationship between bird abundance and habitat parameters on other plots. They found that they were not very successful in predicting changes in bird abundance. There were immediate effects on vegetation

Figure 4.10

Populations of American shrub-steppe birds affected by alteration of the habitat (from Wiens and Roten-berry 1985)

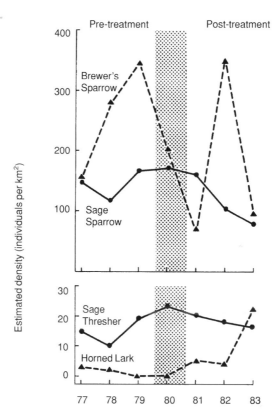

Estimated population densities are shown for four major breeding species in Guano Valley before and after herbicide treatment in 1980 (hatched). Clearly an alteration in the habitat has had an impact on the populations of these species. The herbicide treatment had a negative effect on the population of Brewer's Sparrow immediately after the treatment, but this species rapidly recovered its population level. The Sage Sparrow went into slight decline following the treatment, the Sage Thrasher remained at a similar population level, and the Horned Lark increased somewhat.

but bird numbers changed less than would have been expected. Comparisons of before-and-after results at the study site with those elsewhere showed very little consistent pattern from site to site. As a result, it was hard to tell whether changes were due to the vegetation treatment or whether they might have occurred anyway.

There is a clear moral to this story. A short-term before-and-after study on a single plot could be very misleading. If the study had been conducted in the single years before and after treatment, it would have fundamentally misidentified the effects on Brewer's and Sage Sparrows (Fig. 4.10). The effects that were expected to occur did, but they did not happen immediately and were further masked by large annual variations. The authors suggest that individual birds may have been site faithful in spite of the habitat having become unsuitable.

Such studies clearly need to be replicated and to be conducted over several years. The authors point out that political and funding considerations often demand instant answers. Those from short-term studies may, however, be dangerously misleading, as this study shows.

Summary and points to consider

Transects are particularly suitable in extensive, open and uniform habitats, or areas where there is a low density of birds.

They require a high level of identification skills.

They are the most efficient of all general methods in terms of data gathered per unit effort.

Where their use is appropriate, transects can be more accurate than point counts.

Transects generate less detail than mapping counts.

In addition to walking, transect routes can be followed using ships, planes or cars in appropriate circumstances.

Habitat measurements can be made along sections of the routes to coincide with the division of bird records.

There are four variants of measuring distance to birds: one, two, or several belts and complete measuring.

More elaborate measuring is harder to do in the field but permits more elaborate analysis.

A compromise is required to select an appropriate measuring system for the purpose of the study.

There are no fixed rules for transect counts, but careful thought needs to be given to:

Selection and location of routes
Number of visits
Walking speed
Whether or not to count birds according to activity, age and sex
Distance estimation
Observer and other biases.

5

Point Counts and Point Transects

Introduction

If you stand at one place, it is possible to count all the birds seen and heard. At its simplest, such a method repeated over several places will assemble a list of species present in an area. With some assumptions about how detectability of birds varies with distance, this can be made into a powerful method of measuring relative or absolute abundance (Reynolds *et al.* 1980). The method has come to be widely used for counting songbirds, particularly in France and America. If the habitat is measured around the census point, inferences can also be drawn about habitat selection and preferences of individual bird species or communities (Chapter 11).

This chapter describes the methods used in the field and in calculating densities. The critical assumptions are described with hints on to how to minimise their violation. Several examples are given. One such study provides density estimates for restricted-range lowland birds in Buru, Indonesia. A second example makes an assessment of the conservation status of a rare bird and the third is intended to describe the influence of habitat succession on bird communities.

Point counts are similar in concept and theory to line transects (Chapter 4). They differ in that the distance measured is from the observer to the bird and the counting unit is time, rather than distance. The term point count is conventionally used to describe any fixed radius point count method. Such methods do not take into account differences in the detectability of species. By measuring distances to bird contacts, detectability can be modelled and density estimates adjusted accordingly. New terms, point transect or Variable Circular Plot (VCP) are used to describe these distance sampling methods. Points have the advantage over line transects of being easier to incorporate into a formally designed study. It is easier to locate points randomly or systematically than it is to lay out transect routes because routes require better access, which may bias the habitats sampled. A well spaced sample series of points in an area will provide more representative data than a few transects. Points are often preferred to transects in more fine-grained habitats if identification of habitat determinants of bird communities is an objective of the study. This is because the habitat data can more easily be associated with the occurrence of individual birds (Fig. 5.1).

Figure 5.1

Choosing
transects or
point counts

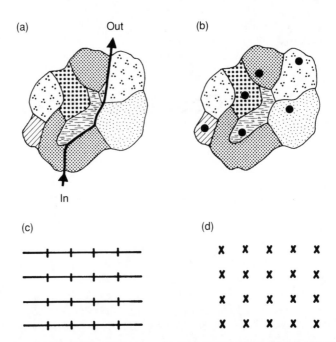

(a) In a fine-grained habitat, such as a wood, a transect following an access route might not be very representative. It would not be easy to divide the bird records into habitat types. Indeed, in this example, two of the habitat types have not been sampled at all.

(b) In the same place, point counts could be set out at random or systematically so as to represent the full range of habitats present in the wood. Each point could also have the habitat recorded around it.

(c) In open country, transects could be set out in a way to cover more of the ground and divided into sections for recording birds and habitats.

(d) The equivalent design for point counts would theoretically record fewer birds but would take about the same time to execute. However, if birds were flushed ahead of the observer, as is generally the case in open ground, this would be a poor design because the observer walking up to a point would scare all the birds away.

The main disadvantage with points is that they are difficult to analyse. The area surveyed is proportional to the square of the distance from the observer. With line transects it is only linearly proportional to lateral distance, with the other dimension coming from the transect length. Density estimates from point transects are therefore more susceptible to error arising from inaccurate distance estimation or from violation of assumptions about moving birds. The routine for analysing point transects using Distance (Thomas *et al.* 1998) requires a rigorous assessment of data and careful manipulation of models to achieve reasonable density estimates.

Points are similar to line transects in requiring a high level of observer skill. By waiting at each point, there is slightly more time to detect and identify difficult birds than in transects. On the other hand, a relatively large amount of time is spent travelling between survey points, whereas data would be collected almost constantly using line transects. In some habitats, there is also the advantage of being able to concentrate on birds without the noise and distraction of avoiding obstacles while walking. In scrub or woodland, points may be preferred for these reasons. Point counts are not as efficient as line transects in detecting numbers of birds. For example, transects offer a chance to record fleeing birds ahead of the observer. At the same time, bias arising from disturbance by the observer is more of a concern in point counts. Every effort needs to be made to reduce such disturbance and this is one reason why point counts are not commonly used in open habitats or for many larger birds where problems of fleeing from the observer are most severe.

Like line transects, point methods are much more efficient than mapping censuses (Fig. 5.2). In a single morning one observer might visit 10 points. If the breeding season lasts for about 50 days and points are visited twice then some 200 points can be studied (assuming that the weather is unsuitable on 10 days). In wooded British habitats this would amount to about 2500 records of birds. In the same time it would be possible to conduct mapping censuses on four plots visiting each one ten times. In comparable habitats, this might generate about 500 clusters or territories. So for the same effort, point counts might generate five times more independent bird data for analysis.

Standardised point methods may exist for certain species or species groups. It is worth considering finding out if a method already exists, before mounting a survey on any particular species. The advantage would be that data could then be directly related to other studies. On the other hand, standardised methods might sometimes be undesirable since different circumstances might be better studied by different designs. Important variables, discussed below, include the number of visits to each point, the measuring of distance to records and the duration of counts.

What do densities derived using point count methods mean?

Very few studies have attempted a comparison between absolute (true) bird numbers and those derived from various different census methods. This is because absolute numbers are extremely difficult or expensive to determine. Many studies have compared results from different census methods, often with the dubious assumption that territory mapping is the most accurate method.

DeSante (1981) conducted a study in Californian scrubland of breeding birds using colour-marking and nest-finding, and then compared the results with densities derived from variable circular plots (VCP). The point counts underestimated densities but most results were within about 30% of the 'truth'. Densities were likely to be overestimated where the species was scarcer and underestimated where it was more common. This is probably

Figure 5.2

Three
different
methods for
a breeding
season
census in a
hypothetical
set of 20 ha
oak woods in
England

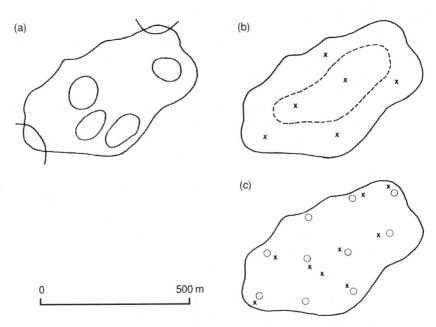

0   500 m

(a) A mapping census. About 150 territories would be mapped for all species. Ten visits would be needed which would take ten mornings. One person could count four such plots in a season. The records would provide the most detailed descriptions of the birds of these sites with most of the species present at least being recorded. Territories would be mapped which would allow comparison with a map of management history or habitat features.

(b) Transect counts. A route of 1 km following existing paths would take about 1 hour to record once. With two visits per route, one observer could record about 40 such routes in the summer assuming they were close enough to allow the coverage of two per morning. In one wood, about 150 bird records (x) might be generated. A longer transect could be fitted into the wood if straight lines were used. This might be difficult to do if the ground vegetation was thick. It could double the number of records in the wood but take so long that it would not be possible to count two such woods in one morning. Habitat records could be taken but it might be difficult to divide the route and the bird records sufficiently finely to reveal the influence of fine-grained variation on birds.

(c) Point counts. Ten point counts (o) would be selected randomly with none allowed to be closer than 150 m from its neighbour. They would take a morning to count including the measurement of habitat features in a 25 m circle round them. Visiting each twice would allow 20 such woods to be counted by one person in a summer. Each would generate about 120 bird records. The description of one site would be quicker but less complete than that from mapping.

If data from many woods are required by the study design, mapping would be too slow. Although collecting slightly less data on birds than transects, point counts might be preferred if habitat associations were also to be investigated, or if there were problems with access, making transects difficult to execute.

because birds in sparser populated regions have larger ranges and make bigger movements around them, so are more likely to be detected. DeSante (1986) has undertaken a similar study in Sierran subalpine forests in America.

Field methods

1. Selection of points

Points to be counted can be laid out systematically or selected randomly within the study area (Fig. 5.3). This allows general inferences to be drawn about the area sampled. The sample design may be stratified. One might, for instance, want to study the effect of succession on the bird community in a certain kind of woodland. An appropriate design might be to allocate 20 points randomly in each of five groups of stands representing different stages of succession. Such a stratified approach would be more efficient for such a purpose than simply allocating points randomly or systematically in a large area which might mean no points fall within scarcer habitat features of interest.

Counting the same bird from two points does not violate any of the assumptions of distance sampling. However, if a bird is counted twice from the same point, this would result in an inflated density estimate.

It is difficult to fit many points into a small area. For this reason, point count methods do not lend themselves to describing the birds, or measuring year-to-year changes in a small area. Nor are they an efficient way of estimating the density of very small populations. In a 20-ha wood, all the songbirds might instead be counted by a mapping survey. If, on the other hand, the study was about habitat factors affecting woodland birds in a large area, points might be more suitable. Several woods would be sampled and habitat data collected at the study points (Chapter 11).

In small patches of habitat, thought needs to be given to the inclusion of points near the edges. These will include birds living in adjoining habitats, so open-country birds might appear in a woodland bird census. One possible way of dealing with this would be to have semi-circular plots where birds are recorded only inside the plot boundary. Depending on the purpose of the study, this may or may not be desirable.

2. Duration of counts

Counts can begin as soon as the observer reaches the point or can be delayed for a few minutes to allow birds to settle down from any distur-bance caused by the observer's arrival. Habitat features may be described or measured during this period, although obviously it would not be a good idea to use this time for any habitat measuring that involved the observer walking around the area.

The duration of counts is an important consideration for any study. The time should be just long enough to detect all the birds at the site, before they move significantly. The longer one stays at a point, the more birds are detected (Fig. 5.4). Normally, however, the majority of records are

Figure 5.3

Systematic,
random and
stratified
allocation of
points

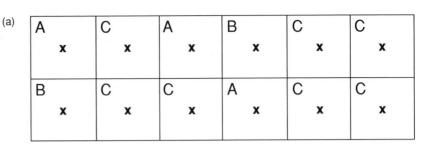

The birds of the rectangular area are best characterised by a random or a systematic allocation of points.

(a) If allocated systematically, three points fall in habitat A, two in B and seven in C which is the most abundant. If allocated randomly, it would also be likely that there would be more points in the common habitat than in the rarer ones but the numbers would probably not be exactly 3, 2 and 7.

(b) A stratified random sample would be better for describing the effects of the three habitats. Equal effort (four points shown) now falls in each habitat. The points were selected by choosing the location of each with random numbers but were constrained so that four fell in each habitat type. In practice, several randomly generated points were rejected after four had been chosen in C but two more were still needed in A. They were also constrained so that no point could lie closer than 50 m from a previously chosen one. Again this was done by rejecting a newly chosen point if it was not suitable for this reason.

Random numbers always look odd but they provide the soundest way of making unbiased statements about the bird communities of the three habitats. Note for instance that by chance one block of habitat C has two points in it while by chance four have none.

accumulated quickly and fewer and fewer are detected in each successive time interval. In a long count, it becomes ever harder to be certain that a 'new' detection is not in fact a bird that was seen some minutes previously and has moved. Longer durations are also more likely to record additional birds moving onto the plot, which contravenes a critical assumption of the distance sampling methods. Some studies have derived density estimates up to ten times reality due to having count durations that were too long. Movement of birds may also bias relative density estimates, owing to differences between habitats.

Figure 5.4

Cumulative
percentages
of total indi-
vidual counts
with increas-
ing count
duration for
three
Hawaiian
species
(from Scott
and Ramsey
1981)

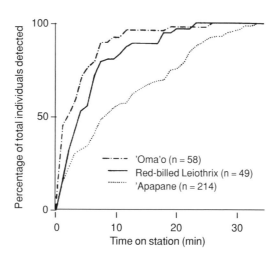

The 'Apapane is the most mobile of the species and the 'Oma'o the least mobile. Half the records of 'Oma'o were achieved in 1 minute while it took 7 minutes to achieve this level for 'Apapane. Longer durations of counts were accumulating records of 'Apapane that had moved during the count which violates a key assumption and leads to an overestimate of density.

Little is known about the details for individual bird species and thus about the practical consequences of such bias.

In most temperate situations, a short count duration of perhaps 5 minutes is to be preferred (Fuller and Langslow 1984). The French IPA (Index d'Abondance Ponctuel) uses 20 minutes. This is probably much longer than is desirable. Longer intervals might be needed, however, in places with a richer bird fauna or with more species that are very hard to detect, such as tropical forests. Time saved by shorter counts can be used to gather data from more points.

3. The recording method
It is possible to separate birds into different types of detection, e.g. males and females. It is common practice to record singing birds separately as it is easier to get good density estimates from them than from more elusive females. This is because the sample number of singing male birds is often higher and that modelling the detection function is easiest if the curve is not 'spiked'. For reliable modelling of the detection function, the curve must instead have a 'shoulder' (see Fig. 4.5, Chapter 4). In Finland, pairs are the counting units. Pairs can be based on a single male or female, a true pair, a flock of fledglings or a nest. If several individuals are encountered the number of pairs is estimated as half the total, rounded up for an odd number. Alternatively it is possible to accumulate all records together. In most circumstances in woody vegetation, more birds are detected by sound

Figure 5.5

Point counts
within a fixed
radius are
not very
satisfactory

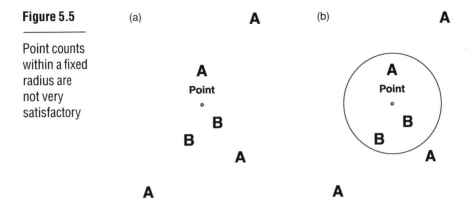

(a) Counts to infinity. Although they actually occur at the same density, more of
bird A than bird B are detected because A is easier to detect than B. This might
be because they sing more loudly or because they are easier to see moving.
As a result, there are distant records of A but B is only detected close to the
observer.

(b) Counts within a fixed radius. The ratio of abundance of the two species has
apparently changed and most records of A have not been used. This could
also be because A is also warier than B and has tended to flee from the
immediate vicinity of the observer.

than by sight. In the breeding season, many will be singing, but at other
times of the year this will not be the case. The rules have to differ by species
according to whether the sexes can be recognised and whether or not song
is frequent. Flying birds should always be recorded separately. They cannot
be treated the same as other records since it is almost impossible accurately
to measure distances to flying birds as they move too quickly and random
movement of birds onto the plot would cause density estimates to be too
high.

4. Distance estimates
With certain assumptions (see below) estimates of density can be made if a
distance measure is associated with each bird detection (Scott *et al.* 1981).
It is always worth doing this and the simplest methods are not too difficult
in the field. The availability of Distance software (Fig. 4.6, Chapter 4)
means that the previously complex analysis of point transect data can be
carried out quickly and efficiently. Distance sampling is further explained
in Chapter 4, and analysis of point transects using the new Windows version
of Distance is described in Fig. 4.6.

The simplest bird counts with no estimation of distance produce results
biased in favour of conspicuous species. Imagine two equally abundant
birds. Bird A has a very loud and carrying song; one of the thrushes would
be a good example. Bird B has a very quiet song, such as a Goldcrest or
Treecreeper. It is obvious that more of bird A will be detected than of bird

Figure 5.6

The analysis
of point
transect data
using
Distance
software

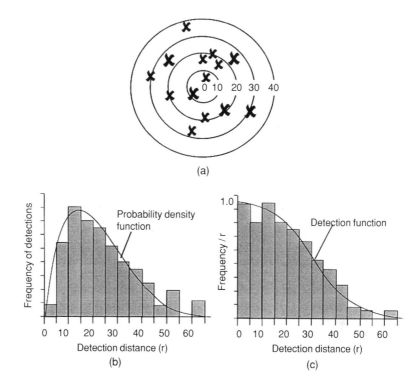

(a)

(b)

(c)

To fully understand this section, the reader should be familiar with the concept of detection functions outlined in Chapter 4, as well as the application of Distance software to line transect data (Fig. 4.5).

The same general rules apply to the analysis of point count and line transect data using Distance software. This includes the detection curve having a 'shoulder' and the elimination of outlying observations which could affect the model fit of the detection function (Fig. 4.5). There are, however, intrinsic differences. The approach to modelling, the importance of choosing the best model and considerations of cluster bias and bird movement are more of an issue in point counts.

Example (b) shows the probability density function for single objects (this may be either single birds or single groups/clusters). That is, the frequency of records in each distance interval is recorded and represented as a histogram. In point transects, distance intervals are arranged in concentric circles (a). This makes it impossible to fit a detection function straight to bird numbers, since more birds will be present at greater distances. A further complication is that at near distances, few birds are recorded. Graph (c) shows how the detection function fits the data when frequencies are transformed by dividing them by detection distance.

Because it is common for no birds to be recorded at $r = 0$ and there are few distances recorded close to zero, the model (c) does not seem to fit data in the first three intervals particularly well. Unfortunately model fit relies on the shape of data at zero and close distances, so there is often a lot of variability between

Figure 5.6

(continued)

model types. When using point transects, particular attention should be paid to choosing the most appropriate model, using all the evidence available.

If distances are measured accurately and there appears to be no violation of any of the assumptions of point transect sampling, Buckland et al. (1993) recommend analysing ungrouped distance data. By 'eye-balling' histograms of the data, it is possible to identify particular attributes that may affect the model fit. If distance measurements appear to be 'heaped' at rounded distance measurements (e.g. 10 m, 20 m, 30 m . . .), it is then best to regroup data. Group cut off points should ideally fall between any peaks, for example if there is rounding to the nearest 10 m then cutpoints of 0 m, 15 m, 25 m, may be appropriate. If probability of detection appears to fall steeply as distance increases from zero (b) then this may indicate a failure in the method. For example, the time spent at each point may need to be increased, so that more distant birds are recorded. However, spending too much time at a point, may increase the effect of bias from random movement of birds. It is not possible to identify this from the histogram, as it causes a general increase in frequencies in all distance intervals – those close to the truncation point because of movement in from beyond the truncation point. Regrouping should similarly be applied to ensure that few birds move between distance intervals during the count.

If the birds occur in clusters (flocks), then detectability may be a function of cluster size. In each case, flock size should be noted and the distance recorded as the centre of the cluster. If distances are divided into intervals, the whole cluster is included as long as its centre lies within the interval. Likewise, a cluster would be ignored if its centre fell outside a fixed width transect or point. At small distances both large and small clusters will be equally visible, but at large distance smaller clusters may be less visible, resulting in mostly large clusters being recorded at greater distances. The change in recorded cluster sizes with distance can be shown in a simple scatter plot. There is a facility in Distance to deal with clustered samples – which is applicable to both line and point transect sampling methods – which estimates mean cluster size, allowing for such size bias by regressing cluster size on probability of detection. If birds are recorded in clusters then the sample number of birds generally needs to be larger.

B even though they are actually of equal abundance. This is because bird A can readily be detected at up to 100 m even in a closed wood, but rather few individuals of bird B will be detected at more than about 30–40 m. The philosophy can be extended to differences in the detectability of a species in different habitats. Starlings, for instance, will be easier to detect in a grass field that is grazed than in a field where grass is long. A study into the density of Starlings on a farm with both grazed and ungrazed fields needs to account for different levels of detection in each habitat type. The modelling facility in Distance is 'pooling robust'. This means that, if certain assumptions are met, the model will be flexible enough to reflect the combined effect of detectability in all these habitats and the 'pooled result' will be a reliable estimate of density (see Fig. 4.5).

The most sophisticated method of point counting involves measuring the exact distance from the observer to each registration. If the bird moves, the distance to measure is to the point where it was first detected. It is hard to estimate distance from oneself to a fixed visible point, although this is made easier with the advent of modern laser range-finders. Distance measuring in closed habitats, such as woodland, is made more difficult as the observer often has to judge distances to audible, but not visible contacts. Because distance measurement underpins the entire method, it is worth investing time into practising estimating distances to visible and audible objects before fieldwork starts.

There are compromises between these two extremes. Records can be grouped into one of two or more circular bands of distance from the observer. If two such distance intervals are used, a sensible dividing point is at about 25–30 m. The disadvantage of this method is that it is a trade-off against accurately recording distances to birds and it requires the researcher to assume a generalised model for the detection function. For reasons explained in Fig. 5.6 it is important to pay particular attention to observations near 0 m.

Methods of surveying using point count methods

The main considerations that apply to transects are the same for points, although measurement errors using point transects are more serious, as area changes with distance squared. The algebra of the calculations is therefore also slightly different. The following paragraphs explore three ways of collecting and interpreting data collected from points. The first example is a fixed radius method of point counting. The second method is also fixed radius, although it uses two distance intervals so that the approximate shape of the detectability function can be assumed. The final method is the most sophisticated point transect or Variable Circular Plot (VCP) method.

1. Fixed radius or point counts
Single (or fixed radius) point counts can be made either to infinity or within an arbitrary range, such as 25 m from the observer. Different species are measured on different scales, and abundance across species and across habitats cannot be compared if there are differences in detectability between them. Large and conspicuous species will be overcounted relative to quiet or cryptic ones (Fig. 5.5). The counts are, however, quick and simple to make in the field. If a fixed radius is used, this bias might be less severe but the actual area counted may be small. With a counting radius of 25 m, ten points counted in a field session cover only 2 ha. By comparison, an 8 km transect with a similar distance limit would cover 40 ha. Another problem with fixed radius point counts is that birds might be just beyond the zone included and therefore potential records have to be ignored.

2. Distances grouped into two intervals

It is not possible to test the fit of a variety of models using two distance intervals, instead an assumption is needed about the form of the relationship between distance and detection probability. With distances grouped into two intervals ($0-r$ and r to infinity) relative densities can be corrected for variation in detectability of species. It is then simple to use Distance software to adjust data based on a predetermined model such as half normal.

3. Exact distance measurement

In the past, collecting data grouped into several distance intervals has been recommended as a means to economise on the rigours of collecting exact distance data. Now that analysis has been simplified by modelling software, we recommend that wherever possible, distance is recorded exactly. If distances to all records are measured, then there is considerable advantage in manipulating the data set to remove bias associated with outlying observations and to some extent rounded distance estimates. For example, as the function of detection differs for each species, post-field work 'grouping' into distance intervals may be advantageous. The cut points between groups can be based on the shape of each detection curve and the effect of 'heaping' at certain densities, i.e. rounding of distances at 20 m, 25 m and 50 m can be reduced to improve model fitness. A serious survey should however, consider using survey lasers to accurately measure distances so heaping would not be a problem. The statistics of this approach and the use of Distance software are further explained in Fig. 5.6 and Figs 4.5 and 4.6 (Chapter 4).

Full distance measurement may not be practical or economical in situations where there is a high density of birds. If, however, a project is studying only a single species or a small group of species, it would be advisable to collect the most exact distance estimates possible. Best practice is to collect data that will be most useful when it comes to analysis, without jeopardising the accuracy of fieldwork. This may mean collecting exact distance data for only certain target species or halting exact distance estimation during periods of intense bird activity.

Assumptions

As with all methods, it is important to understand the assumptions that are made. In this way, steps can be taken to ensure that they are met in the field or that unwarranted conclusions are not drawn if the assumptions are violated. Point transects require the following assumptions to be met.

1. Birds do not approach the observer, or flee

This assumption is most conspicuously untrue in open country where very few birds will remain within 10–20 m of an observer. It is also violated by most larger species, which in general are warier of humans. If data are to be grouped into distance intervals, the critical assumption is that fleeing birds do not move from one interval to another. In practice if the number of registrations per unit area is plotted against distance from an observer it is often found that most birds are some tens of metres away with very few remaining close to the observer. This suggests that the closest recording interval used in analysis should be big enough to embrace the abundance of birds that have fled a short distance. A larger interval-width is generally required in more open habitats. Violation of the assumption about fleeing will generally lead to underestimates of density. If birds approach the observer, densities might be overestimated.

2. Birds are 100% detectable at the observer's location

The distance sampling methods described in this book all assume that the bird is fully detectable at the observer's location. In practice this assumption is most likely to be violated by very quiet and skulking species including most nocturnal birds such as owls, which are barely detectable by point count methods (or by any other general census method). It is likely that in high forests birds directly overhead could be missed if there is dense foliage between them and the observer. Violation of this assumption may lead to an underestimate of density. By recording just singing males, it may be possible to derive a more accurate density estimate.

3. Birds do not move from their 'snapshot' location before detection

The mobility of birds is one of the reasons why the number of birds counted increases the longer one stays at a point. Bird movement also causes considerable problems in recognising individuals. Imagine that a bird calls from one direction, goes quiet for a period and moves, and then calls again. Non-responsive movement, and subsequent detection near to the observer is a source of considerable bias in point transect surveys. It is difficult to avoid counting that individual twice. The problems of mobility are best dealt with by using a short count period (5–10 minutes). If the period is too short, however, one may miss the more silent or skulking species nearby. Violation of mobility assumptions make it difficult to compare species such as raptors, pigeons or corvids, with small songbirds. Densities of mobile species will be overestimated. Point transects can also be used in winter but birds are generally then more mobile than in the breeding season so densities might be overestimated. Bird species moving in parties in tropical forests cause similar difficulties.

4. Birds behave independently of one another

Sometimes one bird reveals itself by call or a song as a result of another individual calling or singing (Fig. 5.7). The effect of such behaviour might be to make high densities easier to measure accurately than lower ones. There may also be an interaction between density and mobility of birds. At lower densities, individuals may have larger ranges and move more, thus violating the assumption above. In other words, it would be possible for the number of birds detected to have a non-linear relationship with the number actually present.

Figure 5.7

Birds are
assumed to
behave inde-
pendently of
one another.
This may not
be true

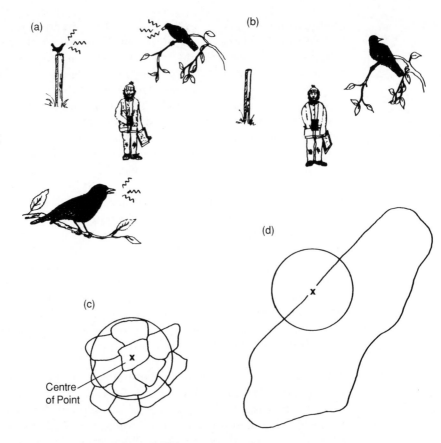

(a) At high densities, individuals may be more vocal as they sing at each other
to defend and advertise their territories. Territories are small (c), and individuals may not move far during the count period.

(b) At low densities, individuals may not have a near neighbour and may be
quieter. The owner of the larger territory (d) may move further during the count
which could lead to an overestimate of density.

**5. Violations of the above assumptions do not interact with habitat or
elements of study design**

It is quite possible that the detectability of birds or the moving behaviour of
birds might vary by time of day or habitat. In this case bias might fall differently in different circumstances. Problems with matters such as time of day
or weather are dealt with by standardising the method to a fixed range of
conditions and ensuring that the analysis compares data derived using the
same method. Interactions with habitat are potentially more serious since
the effects of habitat variation may well be an objective of the study. It is
fairly evident that fleeing behaviour and detectability may vary with
openness of vegetation. If no distance measuring is done then counts in
different habitats will quite clearly be unrelated to densities in any simple

way. If counts have been made with one of the methods that includes distance estimation then it is possible to correct for each different kind of vegetation separately and produce a pooled overall estimate.

6. Distance estimates are accurate

Accurate distance measuring is particularly important for point transects because bias arising from errors varies geometrically with distance. Observers should be trained to assess distances. Training has been shown to have a marked improving effect. If permanent points are counted repeatedly, it is possible to mark some fixed distances to use as reference points. Some studies have used optical range-finders or survey lasers to measure distances. Some studies have recommended that the observer moves in order to locate birds. This has the disadvantage of creating further disturbance and exaggerating the problems of birds fleeing.

If distances are grouped into intervals rather than by full distance measuring methods, it is only necessary that individuals are put in the correct intervals. This is obviously not as difficult to do in the field as complete distance estimating. In the case of the two-interval method, it is necessary only for the observer to have a very good idea of what the single radius looks like and to check each bird as to whether it is within or beyond that distance.

7. Birds are fully and correctly identified

Because most detections are by sound, point count methods require a particularly high level of field skill. The option of moving to see and identify a bird is often not available. Observers, therefore, need to be fully familiar with all the species in the area and with the separation of any that sound similar. Training for consistency in identification is almost certainly worthwhile in an area unfamiliar to the observer. Tapes might be used for training and for assessment of standards of observers. If many observers are used in one study, thought should be given to how they are allocated to particular points. In this way, study design could, for instance, prevent bias arising in the counts from one habitat type because one aberrant observer did all the recording there. If points are visited more than once, observers could be swapped. In this way, it would be possible to check their degree of consistency. Because it is common practice to truncate 5–10% of the most distant bird records, birds that are not identified a long way off can be simply ignored.

Examples of the use of point count methods

1. Densities of restricted range lowland birds of Buru, Indonesia

Marsden *et al.* (1997) aimed to compare the abundance of Buru's restricted range lowland birds in different habitats. Thirteen out of 19 such species were successfully recorded. Habitats were divided into primary forest, secondary forest, 2–5 year old and 12 year old logged forest and non-forest (coastal vegetation and grassland) habitats.

The survey used the Variable Circular Plot (VCP) or point transect method (Reynolds *et al.* 1980; Buckland *et al.* 1993). Compared with other fixed radius point count methods (Hutto *et al.* 1986) it allows for the adjustment of bird density according to detectability. Points were located systematically 250 m apart either along transects or 50 m to one side. All contacts to birds were recorded during 10 minute intervals and their precise distance from the observer estimated. All objects were entered as clusters of birds, i.e. flock size. If a group was only heard, it was assumed that the group size was equal to the average group size recorded visually for that species.

For all species, between 10% and 20% of observations were truncated to remove unwanted outlying observations. For each species, the optimum interval grouping was chosen by modelling a detection function for each 'trial' and choosing the grouping combination that achieved the lowest AIC (Akaike's Information Criterion) statistic using Distance software.

As is common with multiple-species projects, the data were analysed in a number of different ways, dependent on sample size. For species that were commonly recorded, observations from each habitat were analysed independently. This enabled a separate detection function and density estimate to be calculated for each habitat and a pooled density derived using the known proportion of those habitats on Buru. For less common species, a combined detection function was produced for 'forested habitats'. This method assumes that the detection curves for any given species do not differ significantly between forest habitats, i.e. primary and secondary forest. For species recorded on even fewer occasions, detection curves were derived for 'groups' of two or more species. In these cases, the detection function was not species specific and therefore relied on the assumption that every species in a group was similarly detectable.

Density estimates were only possible for ten species in secondary and primary forest habitats, eight species in logged forest, two in forest habitats combined and two in the non-forest category. Coefficients of variation ranged from 11% to 85% and only seven out of 22 estimates resulted in coefficients of variation of less than 25% of the mean. Several small species were recorded too few times for reliable density estimates. For example, the density of Black-tipped Monarch was 213 ± 112 birds/km^2. Marsden *et al.* were not confident that a density greater than 100 birds/km^2 was an accurate estimate, especially as the species was recorded at only six VCP stations after data truncation.

The survey suggests that encounter rates without adjustments for detectability are not necessarily a good indication of relative abundance. For example, the mean number of encounters recorded at each survey station was 0.56 ± 0.08 for White-eyed Imperial Pigeon and just 0.2 ± 0.05 for Flame-breasted Flowerpecker. Despite White-eyed Imperial Pigeon being encountered nearly three times as often, the resulting density of this species was much lower; just 38.3 ± 7.5 birds/km^2 compared with 171 ± 43.4 birds/km^2 for Flame-breasted Flowerpecker.

This project depicts the strengths and the weaknesses of point transect methods of surveying birds. It uses a good theoretical methodology to consider differences in the detectability between species and between species in different habitats. It also shows the importance of collecting data so that it can be utilised in several different ways, depending on the nature of the sample size and the shape of the variation in detection with distance.

Unfortunately, the Buru study was conducted over a very short time period, not nearly long enough to get suitable sample sizes for anything other than the most common species. The sampling strategy also ignored land above 880 m. A significant area of Buru is highland forest over 1000 m and as such, it would be difficult to determine the conservation status of lowland species that also occupy this area, using the study's density estimates. It is therefore possible that the study does not provide a reliable estimate of species' abundance. In time, however, it could be possible to use supplement plots and combine data over several years to produce more useful density estimates.

2. Monitoring breeding birds in the USA

The American Breeding Bird Survey (BBS) is a large programme which has run throughout the USA and Canada since 1965 (Figs 5.8 & 5.9). It is sponsored by the US Fish and Wildlife Service and the Canadian Wildlife Service but draws much of its support from an organised network of amateurs. The American BBS aims to monitor population trends of a wide range of breeding birds (about 230 species). The information generated can contribute to a variety of objectives of differing scales.

Figure 5.8

Trends in House Finch numbers in the United States from the Breeding Bird Survey (from Robbins *et al.* 1986)

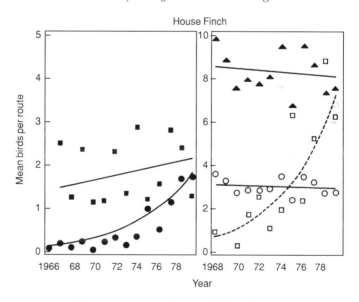

House Finch

The data points are the mean number of birds per census route. The plotted trend lines are calculated in a way too detailed to elaborate here.

Marked increases were noted in the Eastern Region (●) and Southern New England (□) with population levels calculated for Central Region (■), Western Region (◆) and Continental data (○) remaining similar.

Figure 5.9

Mean relative
abundance of
the Plain
Titmouse and
Tufted
Titmouse in
the United
States (from
Robbins *et al.*
1986)

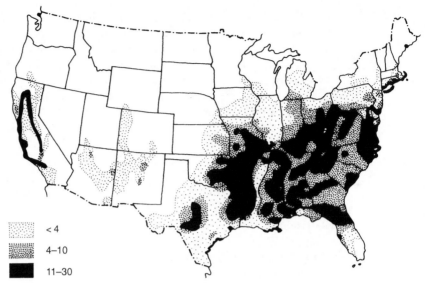

In the Breeding Bird Survey of America, the distribution of routes across the
continent, and standardised methodology, make it possible to draw general
inferences about the geographical variation of abundances of species. This is
not hampered by the abundance figures having no absolute meaning. Note
the tendency which is shown by many species to be more abundant at the
centre of their ranges than at the margins.

1. Measure normal year-to-year fluctuations in numbers of individual
   species.
2. Detect the effects of variations in weather including catastrophic events.
3. Measure long-term trends in numbers.
4. Allow description and analysis of fluctuations and trends on a geographical or habitat basis.
5. Describe widespread biogeographical phenomena such as regional and
   habitat-based differences in relative densities of particular species.

Data are gathered by point counting. Instead of being randomly spaced,
points are instead located along roadside routes each of which consists of
50 points spaced at about 800 m apart (actually 0.5 miles). They are visited
once each summer at a date mainly in June but with some allowance for
latitude. The counting starts half an hour before sunrise and each point is
counted for 3 minutes. The total number of birds of each species is
recorded up to a distance of 400 m.

From year to year, the routes are always run in the same direction. Stop
points for counts are mapped and described once established so that they
are constant from year to year rather than reliant on a car's odometer
(milometer in UK). No methods of attracting or provoking calls from birds

are allowed. The instructions make it very clear that the counts are not expected to be a complete record of all species actually present. Standardisation rather than completeness is essential. Observers are cautioned not to stay longer or otherwise bend the rules in the hope of a 'good' species that they know or suspect should be present.

The American BBS has about 2000 routes which are counted each year generating about 1 500 000 records of around 500 species annually. Each lies within a single one degree block of latitude and longitude and within one state. The majority lie within a single physiographic unit derived from a map of the life zones of America. Organisers try to find observers to count allocated routes. Coverage varies with human population density but this can be allowed for in a stratified design. Further reading can be found in Engstrom and James (1984), Geissler and Noon (1981) and Robbins *et al.* (1986).

Long-term trends were originally calculated on the basis of ratios of counts for a species on the same routes in two adjacent years. Trends are now estimated for each route by a log-linear regression model, which can better cope with occasional missing values. Unfortunately, even though all the count data originating from the American BBS are on a relative rather than an absolute scale, the method is seriously flawed in its design. To accurately monitor a trend in species numbers, the points would need to be randomly spaced with respect to roads. Analysis of trends is impossible from points that are positioned along roads. For example, habitat degradation is likely to be most intense along highways. Over time, the American BBS may derive exaggerated estimates of decline that are dependent on this habitat loss, rather than any real reduction in bird numbers.

3. Estimating the range and abundance of the Azores Bullfinch

The Azores (or San Miguel) Bullfinch is a very rare bird which went unseen for 40 years, was recently rediscovered, and is now known from a very small area of one island in the Azores. The terrain is steep, susceptible to land-slip and covered with impenetrable vegetation. The objective of a 20-day study was to assess the Bullfinch's range and numbers to start making a conservation plan. Lack of time, lack of prior knowledge and the complete inaccessibility of so much ground made much formality of method difficult.

Point transects were chosen as the census method. A mapping census would have been completely impossible because of access problems and it was not known where to locate plots because of a lack of prior knowledge. Transects were considered but access was again a problem. The bird was expected to be very quiet like the European Bullfinch and it was thought that even the noise of walking over difficult ground might lower the chances of detection. The points were spread at intervals of 200 m along any access routes found in or near the suspected range of the species. Counts lasted for 10 minutes since the bird was expected to be quiet and inconspicuous and possibly able to remain close to an observer, but undetected, for several minutes. This was checked by separating records into two 5-minute periods. Records were allocated to within or beyond 30 m at first detection. Descriptive records of the vegetation were made at each point so that a comparison could be made between points with and without Bullfinches.

The method was described in terms of numbers of points covered and their locations, the time of counts (10 minutes) and the radius separating near and far records (30 m). This survey could be repeated at any time and tell whether this very rare and localised bird is declining or increasing in numbers.

The results are published elsewhere (Bibby and Charlton 1991). It was possible to narrow down the likely range of the bird because points with and without the bird were recorded. General observations on habitat were possible for the same reason; habitat was described where the bird was found and where it was not. The counts could be turned into relative density estimates by assuming a relationship between detectability and distance. By applying such an analysis to areas in which we found the bird, we could make a first estimate of densities and total area of the likely range and thus of total numbers. The total population estimate of about 100 pairs should be treated as little better than an order of magnitude estimate. The areas that were accessible were not the same as those that were not so it was possible that the sample was not representative of bird density across its very small range (about 500 ha). Because counting methods were standardised, it can, however, be claimed that changes of numbers will be detectable when the same method is repeated in the future.

This was not a perfectly designed study because access did not allow it. This case illustrates that some consideration of methods is none the less worthwhile. Many of the world's endangered birds (Collar *et al.* 1994) have never been counted, which is a severe obstacle to their conservation (Green and Hirons 1988). Even partly designed and formalised methods are more useful than casual records in the process of assessing changes of numbers and threats.

4. Studying the effects of vegetation in young forestry plantations

The aim of this study (Bibby *et al.* 1985) was to investigate the factors influencing bird occurrence in young forestry stands at the start of the second rotation. There was only one summer in which to do the work with two observers. It was important to obtain results of wide generality so all the forests in North Wales were included. Stock maps showed all possible plots below 10 years of age. The only way to collect enough data was to use point transects. Size of area cut was potentially important so study plots were chosen (on a random basis) to cover the range of sizes within the six main forest blocks. The number of points per plot was set to be proportional to size of plot, and individual locations were selected randomly with the constraint that two sample points had to be at least 60 m apart.

Each point was visited twice at an interval of about 30 days because many of the birds were migrants with later breeding seasons than the residents. The two observers were trained in distance estimation and checked that they were familiar with all the likely birds. Further, to remove any observer bias, each observer made just one of the two visits to each point. Vegetation measurements were made at each of the sampled points. Many of the study plots were densely vegetated and difficult to access but 326 points were visited in a 2-month field season. Over 3200 records were obtained of 31 bird species.

The study would have been very difficult to execute by any other counting method. An advantage of point transects is that one can compare vegetation features at points with or without particular birds (Fig. 5.10). A problem with the method is that estimates of relative densities cannot readily be compared with much other published data. Mapping results have been published more extensively to date.

Figure 5.10

Habitat selection by the Chiffchaff in conifer plantations in Wales (UK) (from Bibby *et al.* 1985)

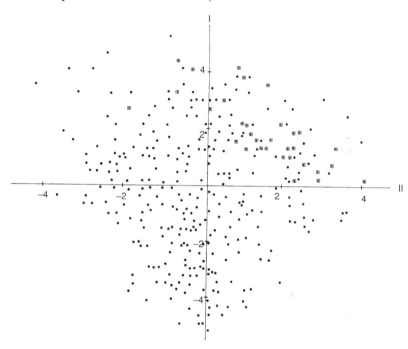

A total of 326 point counts were conducted and habitats measured at the same places. This allowed a variety of analytical approaches to describing habitat selection of the bird community.

In this figure, all the census points are plotted according to their position in a two-dimensional view of habitat. The axes (derived from principal components analysis) approximately coincide to age (Axis I) and conifer/broadleaf mixture (Axis II). Chiffchaffs (marked with squares) occur only in forest stands with comparatively mature vegetation and a high broadleaf content.

Summary and points to consider

Point count methods are suitable for conspicuous birds in woody or scrubby habitats.

They are suitable for study of extensive areas but do not provide the level of detail of mapped counts.

They are more efficient in terms of data collected per unit effort than mapped censuses but less so than transects.

They may be more appropriate than transects in areas where access is poor or where habitat is very fine-grained.

In open country or large-scale habitats with good access, transects may be more suitable.

Point count methods have special value in habitat studies when habitat is measured at the counting points.

Data collected from points can be analysed using Distance software.

There are no fixed rules for point count methods but careful thought needs to be given to:

Selection and location of points

Number of visits

Duration of counts

Measuring of distance to records

Reducing observer bias through training.

6

Relative Measures for Bird Communities in Habitats with High Species Richness

Introduction

The traditional bird counting methods developed in temperate latitudes do not always work well in the tropics. A variety of problems conspire together and reach their very worst in tropical, high canopy, evergreen forests. These are some of the most species-rich habitats on Earth and there are pressing conservation reasons to know more about their variety and the impact of Man on them. It is worth understanding the causes of some of the problems involved. The methodological challenge is to get round them.

1. Large number of species

The large number of species means that very high levels of observer skill are required. In some lowland forests in South America there may be as many as 500 species in the forest or associated with features such as rivers (Terborgh *et al.* 1990). In many countries these are not described by adequate field-guides and the calls are not all documented. Very few ornithologists have the experience to deal with these circumstances and it is time consuming to learn to be good enough.

2. Many species at low densities

In part this is a result of the very large number of species. As a result, a very large effort would be required to generate adequate sample sizes for most species. It is common in a day in the field to find that many of the species are only encountered once (or not at all) and only for very few do reasonable sample sizes accumulate.

3. Unknown seasonality

Most north temperate birds sing and breed in the period February/March to June/July and fairly predictably according to species and location. In the tropics, breeding can be spread over the whole year. With long-lived species, only a proportion of the population of a species may breed in any one year. Breeding may also be triggered by different stimuli than in temperate areas and thus the onset of the rainy season (for example) may be highly important. Start dates for breeding may vary by weeks between years (or even fail completely in some years), so comparing results between sites and seasons is problematic.

4. Varying lifestyles

Bird counting in tropical regions must also take much more consideration of the widely varying lifestyles and behaviour of the birds, which show far greater diversity than in the Holarctic where the standard bird census methods have been developed. In the Neotropics, for example, both the males and the females often sing, across a wide range of different bird groups. In groups such as tinamous the sexes are alike, forage apart, there are several species in many areas, and they probably both sing. In the tropics there are also far more social or cooperative breeders than in temperate areas, and non-territorial species are also present (e.g. trap-line feeding hummingbirds) which further complicate censuses. The ecology and abundance of a number of species are also linked closely to the behaviour of other species. Examples are ant- and monkey-following birds, honeyguides, brood parasites, etc. Some birds are also found almost always either as members of mixed flocks, or within single species flocks (e.g. parrots, toucans and certain icterids).

5. Low calling rates

Some species such as woodcreepers or tyrant-flycatchers vocalise only rarely, perhaps only once or twice per day and often at a very specific time, even before it is light in the morning. Other species are almost non-vocal and are very secretive (e.g. some forest rails and pheasants). They will almost always be missed by the standard methods.

6. Dense vegetation and difficult access

In steep or wet areas with tangled vegetation it can be physically very difficult to get around at all. Concentration on personal safety makes it hard to pay attention to birds as well, while noise may scare elusive birds away. Many tropical species have very small ranges, often in places a long way from easy access. Tall canopies up to 50 m high mean that it is difficult to get close to some species either to hear or to see them.

The combination of these difficulties may mean that the standard methods do not work effectively. Clearing census routes for transects or mapping may take a huge effort. Walking without a cleared route can be so noisy and difficult that counting birds is unrealistic. Selecting points at random is a problem when chosen sites cannot safely be accessed. Measuring distances to calls of unknown volume is virtually impossible. Most species will only be heard and the identification of several hundred species by call, especially within the short time demanded by point counts, is beyond the field ability of most people. Results are likely to be highly observer biased with less experienced people recording a larger proportion of the loud and frequently singing species with distinctive calls and missing more of the quieter, less frequent and less distinct cues.

The mapping method can be used in tropical forests (e.g. Terborgh *et al.* 1990; Moyer 1993), but it is extremely time consuming to mark out the plot and then to collect sufficient data to identify the clusters. The mapping method also relies on being able to identify territories. This may be hard in areas of the tropics if all species cannot be relied on to be breeding at any season or even in any one year and some species may not defend conventional territories. Very few studies in tropical rainforest regions around the world have tried to produce a complete enumeration of the bird species present and their densities (Bell 1982; Brosset and Erard 1986; Brosset 1990; Terborgh *et al.* 1990). These studies typically involve the detailed study of a single plot of forest, ranging in area from 2.5 ha to 97 ha, over a period of weeks, months, or even years.

The most comprehensive of these studies has been that of Terborgh *et al.* (1990), which was conducted in the Amazonian rainforests. This study worked in a single fairly large (97 ha) plot and attempted to find all the species of birds within that plot and assess their abundance. The large plot allowed even those species with the largest territories to be present in the area. Over a period of years prior to the main study (which lasted 2 months) the authors learned the calls of 500+ species of birds, and the calls were used as the main means to census the birds. There has been no comparable study since, which is a comment on the skill and effort required.

Rationale for some simpler methods

The acute difficulty of conducting formal censuses in some of the circumstances described leaves two options, disregarding the possible third of giving up. A more positive alternative might be to limit study to fewer selected species for which it would be possible to use rigorous methods. There are, however, many reasons for wanting to conduct avifaunal studies in poorly known but species rich tropical areas. There are less rigorous, but nonetheless often sufficient ways to do this.

Characterising areas by their variety of species, the occurrence of individual species and possibly the relative abundance of species may be sufficient for many study purposes. Both in theoretical studies and for applied questions about resource use and conservation, such data may well be sufficient and certainly better than the current gaps in our knowledge of tropical avifaunas. Relative abundance is a concept that greatly confuses real abundance and detectability. Many apparently scarce birds may not be numerically scarce at all, they could simply be hard to find. Some apparently locally abundant species might actually be quite scarce; they just impose themselves on the counter through being noisy or conspicuous. Relative abundance measures are certainly not comparable across species. Between sites or habitats they are more comparable within species with some caution for differing effects of conspicuousness in different habitats. Aggregated measures of bird communities, such as species richness, are easier to measure in a comparable way.

A class of methods described in this chapter consists of three elements. The first is to use the best possible efforts to find species in circumstances where this is not easy. The second is to record and use some measure of effort to standardise results. Thirdly, but optionally, data may be compressed into an aggregate measure of species richness or community diversity. These three elements are described in turn.

Finding birds

There are a variety of field methods to improve the number of species found in an area. They amount to no more than the deployment of good fieldcraft and would come as no surprise to an experienced ornithologist, but a checklist may help.

1. Prior knowledge and expectation

It is a great help in finding species to know what might occur in the area and have a list of possible species that will be particularly difficult or valuable to find. These might especially include restricted-range species endemic to the area (see Stattersfield *et al.* 1998) or threatened species (see Collar *et al.* 1994). Regional or national faunal lists and reviews will be useful sources.

2. Learning the birds

There is no substitute for being able to identify most birds seen or, more usually, heard, with speed, reliability and confidence. Field-guides are some help, if they exist. Calls have largely to be learned in the field or from existing recordings before leaving home. In many genera or families there can be several species rather similar in appearance or voice that could occur in any one area. It is particularly valuable to appreciate what these might be and know how they could be separated. Time with an expert at the start of a study is invaluable.

3. Indigenous and local knowledge

While unlikely to be formally schooled in ornithology, local people may know a lot about the habits of birds, especially if they are hunted or have other value to humans. Knowledge can be obtained by informal discussion or assisted by asking about pictures. The desire to please in answering can take the form of suggesting the presence of more species than are actually known. This bias can be assessed by asking about species that would not occur anywhere near. Local assertion of the presence of a species may not be taken as an absolute record but indicates the value of looking harder and local people may know where, when or how best to look.

4. Time of day and season

The seasonality of many tropical species is unknown. Different species may breed at different seasons or differently in relation to unpredictable seasonality such as rainfall. Other species might be short-distance, long-distance, altitudinal or erratic migrants. It is thus difficult to indicate a definitive season for counting in any one area and the coverage of a greater part of the year will often reveal more species. The most striking feature of diurnal variation is the extent to which some species which call very rarely do so before it is even light at dawn, so this is an important time to be in the field, ideally not always in the same place.

5. Micro habitats
In species rich areas, many birds are very discriminating in their habitat requirements. In seeking to find all species, it is thus important to find and give attention to all such habitats. Examples might include stream-sides or bamboo thickets.

6. Altitude
Altitude is a strong determinant of the occurrence of individual species that may characteristically appear or cease to occur regularly within a few hundred metres of altitude. Many species pairs replace each other by altitude. In general, species richness falls with height but endemism may rise and many restricted-range species occur at greater altitudes. It is thus important to work all altitudes within a study area and to be aware of the possible effect of altitude in making comparisons between different plots or areas.

7. Fruit, flowers
Some species may be attracted to localised resources, such as fruiting trees or particularly attractive flowers. Such places are well worth revisiting or waiting at to collect fuller species lists.

8. Vantage points
High canopy species are often hard to pick up and the best chance may come from watching from an appropriate vantage point on steep ground giving views into the higher levels of trees below. Other species such as raptors, parrots or other frugivores may be picked up flying above the canopy and are worth watching for from vantage points offering a good vista over the trees. Some species are particularly mobile at dusk as they go to roost.

9. Flocks
Many tropical species habitually occur in multi-species flocks and still more will temporarily join flocks passing through their ranges. Some flocks, especially in the neotropics, are particularly associated with ants. Flocks are always worth careful attention; follow them if possible, to pick up extra species.

10. Mist nets
In taller forest habitats, mist-netting studies may capture no more than 40% of the species present, even when the sampling effort is very large. Mist netting is not regarded as useful for conducting a census of the birds present in an area, nor of the population density of different species (see Remsen 1994). Setting up the nets and manning them takes a great deal of time and effort. However, mist-netting can be useful to locate shy under-storey species which might otherwise be missed in a more general survey

programme, and is particularly useful for inexperienced teams as it allows birds to be studied in much greater detail before identifications are made.

11. Tape recording and playback

Many species can be induced to call or to approach a tape recording (of themselves or their species) played to them (e.g. Parker 1991). Tapes may be acquired from specialist sound libraries or ornithologists who have worked in the same area. This has an advantage that their identity will be known and is a way to search for species that may possibly occur in the area. Alternatively, unknown calls can be taped on the spot using a muffled directional microphone and played back immediately. Skulking species can thus be induced to show themselves.

Controlling effort

The effort by which species lists are collected can be measured by time or by length of list (Table 6.1). Such studies can be conducted in ill-defined areas or in quadrats of fixed size. Derived measures for individual species are frequency of occurrence in the lists. These are not true abundances and certainly cannot legitimately be compared across species. They might more plausibly be compared between species across sites. Community measures of species richness or diversity can be derived from these data.

1. Encounter rates

Transect or point counting are easier without measuring distances to individual birds. The derived results can be summarised into relative measures of individuals per point, per kilometre or per hour. These are not very

Table 6.1

Relative methods consist of data collecting with some measure of effort. The data may be as simple as lists of species which occur

Field method	Summary statistics for species	Derived statistics for bird community
Point or transect counts with no distance measuring	Individuals per point, per km walked or per hour walked	Total species richness
Species lists per time unit (hour or day)	Frequency of species per sampling unit of time	Species diversity index
X species lists (McKinnon method)	Frequency of occurrence of species per list	
All the above constrained by area	Any of the above for a fixed area	Either of the above for a fixed area

satisfactory approaches. They fail to deal with the circumstances which make formal methods difficult and they fail to exploit the power of formal methods.

An unusual point counting method has been proposed by Parker (1991) and Remsen (1994). The observer simply locates a suitable study area and then uses a tape recorder to tape the pre-dawn and dawn chorus, and the evening chorus. Selected recordings of any species believed to be different can also be made during the day and at night time.

The tapes made are then carefully labelled and can either be listened to during the day, or taken back to the office where they can be compared with the songs of known species. Through this method it is possible to obtain an almost complete assessment of the list of species in the area in a very short period of time. Moreover, there are some possibilities for assessing the abundance of different species, particularly if several tapes can be made from areas separated to such an extent that the same bird calls cannot be heard on each tape.

If the observer is highly skilled in bird voices then this can be a powerful method. However, the level of skill demanded is very high and even with a good library of bird calls interpreting the tapes can be a major challenge, especially in areas where there may potentially be several hundred species calling. Data could obviously be improved by allowing a larger sample of skilled ornithologists to listen to the tapes. Some progress has been made with amphibians in automating the process of identification, in South Africa for example. An obvious constraint of this method is that it will not record species which remain inaudible.

2. Timed lists

The idea of timed lists is that all species recorded within a fixed time are listed. Lists are collected repeatedly and the relative abundance measure, at its most simple, is the frequency of occurrence of a species on all lists. To make this possible a reasonable number of lists, say a minimum of about 15, are required. For studies at a relatively fine scale, lists might be collected in one hour periods. For larger scale studies, day lists might be more appropriate.

The Timed Species Count has been developed in tropical Africa to gather comparable data on the bird faunas of sites ranging from forest through to savannah habitats. The data permit between-site comparisons, which can be repeated in the future to see how the fauna and population levels of constituent species have changed (see Pomeroy (1992) for additional details).

The method does not really require a study plot, although study sites of around 1 km^2 are recommended and these could be regarded as plots. The area should be standardised as far as possible and the habitat, weather, etc. should be recorded and also taken into consideration when results are assessed. Sites selected should be relatively homogeneous and typical of the area as a whole. Thus a small swamp in a large woodland area should not be included in a site. Ideally all sites would be randomly selected, but the choice is often restricted by problems of accessibility.

	Minutes	Site and weather	Date and counters	Time and number
Table 6.2		Ruhizha	27 July 1989	0830–0930
An example of the field record of a		Fine and sunny, 18°C	Counters: TB and DEP	06
Timed Species Count (TSC), made in an open part of	10	Masked Apalis Chubb's Cisticola Chestnut-throated Apalis Regal Sunbird		
Bwindi Forest, south-western Uganda at	20	Black-headed Oriole Luhder's Bush Shrike Sunbird sp. Bronze? Black Roughwing		
approx. 2200 m altitude (after Pomeroy 1992).	30	Yellow-fronted Canary Yellow-whiskered Greenbul Olive Pigeon Blue-headed Sunbird		
	40	Common Bulbul Yellow-rumped Tinkerbird		
	50	Angola Swallow White-necked Raven Yellow White-eye		
	60	Northern Puffback White-breasted Crombec Black Kite	TOTAL = 20 species	

The Timed Species Count consists of a simple list of the birds (see Table 6.2), in which all species that are positively identified are listed, in the order seen or heard within the period of one hour. Other lengths of time for recording have been used but a deviation from the standard period will make comparisons between studies more problematic.

To make the TSC count the observer moves slowly through the study site, listing any species that occur anywhere within it, regardless of how far away. Species flying over are included if they are 'using' the area, for instance swallows feeding, kites looking for food, or raptors displaying. At each site a series of TSCs is made – at least 10 and preferably 15 or 20. These counts are ideally made at various times of day and at different seasons in order to give a full estimate of the species that are found in the site through the year.

Various methods can be used to derive indices of relative abundance from the raw data. The simplest is the frequency of occurrence of each

Table 6.3

Analysis of Timed Species Counts from Ruhizha, Bwindi Forest, Uganda (after Pomeroy 1992)

B	F		01	02	03	04	05	06	Mean
135	F	Crowned Eagle				1			0.2
138		Black Kite						1	0.2
339	F	Olive Pigeon	5	3	6	3	6	4	4.5
372		Great Blue Turaco	1	2	5	5			2.2
459		Speckled Mousebird				1			0.2
488		Cinnamon-chested Bee-eater	4	1					0.8
503	F	White-headed Wood Hoopoe	6			6			2.0
513		Black and White Casqued Hornbill					2		0.3
515		Crowned Hornbill	5						0.8
533		Grey-throated Barbet			1				0.2
548		Yellow-rumped Tinkerbird	2	1		6		3	2.0
556	F	Yellow-billed Barbet			1	1			0.3
584	F	Fine-banded Woodpecker				5			0.8
640		Black Roughwing	4	6		4	3	4	3.8
649		Black-headed Oriole	6	6	6	5	5	5	5.5
653		White-necked Raven						2	0.3
671	F	African Hill Babbler		1					0.2
676	F	Mountain Illadopsis					5		0.8

Only six counts are shown, including the one in Table 6.2, and only the first 18 species are shown. The species are arranged taxonomically according to the standard East African list. The first column (B) gives species codes according to this standard list. The F column indicates where the bird is regarded as a forest specialist, and therefore of greater conservation significance. Species are scored 6, 5, 4, 3, 2 or 1 according to the ten-minute period in which they are recorded.

Note that the mean score for each species is here calculated on the first six TSC counts. Normally at least 15 counts are made at any one site, at various times of day and at different seasons. Even after six counts it is apparent that there are a few common species and many rarer ones, including most of the forest specialists.

species. A more elaborate method is described by Pomeroy (1992), see Table 6.3. Field data from individual Timed Species Counts are rank-ordered according to 10-minute recording periods within the 60-minute TSC. For each of the lists, species are scored 6 if seen in the first 10 minutes of the count, 5 when seen between 10–20 minutes and so on until the score is 1 for species seen in the last 10 minutes. The idea is that the most abundant species will always be seen sooner than the scarcer ones, though it is not really clear whether there is any good theoretical basis for the weights used. The simpler species frequency approach has the inherent advantage of making fewer further assumptions.

The length of the recording period can be altered if desired, but if comparisons between sites are sought then these periods should be the same, and it is recommended that the 60-minute recording period should be used in most cases.

In the above it is suggested that counts are undertaken throughout the year and at different periods of the day. This may not be practical for many types of study and comparable data could still be gathered if the recording period (time of day, or season) were standardised such that between-site comparisons could still be made.

3. X species (McKinnon) lists

This method was proposed for use in South East Asia by McKinnon and Phillips (1993). It is a highly pragmatic method; so simple that it can be used by lots of different people in different places, and robust enough to produce species lists for many different sites, which can be compared. In essence it records species on fixed length lists rather than in fixed time periods. The idea is that this in part compensates for the fact that less skilled observers will take longer to accumulate species lists. If time is the effort measure, this will be important.

This method does not require a study plot. The observer walks slowly around the study area, using paths where available and through open areas of closed forest where paths are not available, covering the area as thoroughly as possible. The whole day can be used to collect species-lists and thus the method is extremely time-efficient. Details of the study area should be recorded (altitude, habitat, aspect, weather, etc.) to allow analysis after the work has been completed. Some attempt at standardisation of the area studied (in hectares) would also be strongly advisable to permit valid inter site comparisons.

The field method is to list the first 20 (or another number might be chosen, see below) birds encountered (seen or heard) without any limit on the time taken to compile this list. It is important that names are applied to all the species encountered, whether or not they can be identified on the first occasion, they can be named later and as experience develops fewer and fewer species will lack a 'real' name. The first list will contain 20 new species. When the first list is complete a second one is started until that also contains 20 species, of which some will be repeated from the first list (Table 6.4). This process is continued until at least 10 lists have been produced for the study site. The number of new species recorded in the later lists falls until few new species are being added to the total. If the new species in each list are marked with an asterisk, they are easier to count when the accumulation curve of new species is eventually made. When the accumulation list reaches a plateau then the regional avifauna has been adequately sampled. The relative abundance measure for each species is the proportion of lists it occurs in.

Tests of the method by Poulsen *et al.* (1997), in Andean cloud forests,

Table 6.4	List	I		II		III		IV	
Example of a computer database produced using running lists (10 species lists in this case) of birds from the Udzungwa Mountain forests, Tanzania		1	**s.mon**	1	**p.pla**	1	s.cap	1	p.ste
		2	**t.liv**	2	a.mil		s.cap	2	**p.fla**
		3	**p.ruf**	3	**a.oro**	2	s.mon	3	a.mil
			t.liv	4	n.oli	3	**b.bre**	4	**m.sti**
		4	**p.bil**	5	**a.ful**	4	**p.ste**	5	**c.col**
		5	**a.ful**	6	**z.sen**		b.bre	6	a.ful
		6	**n.oli**	7	**c.del**	5	**x.udz**	7	**e.alb**
		7	**a.mil**	8	**s.cap**	6	a.mil	8	**b.mar**
		8	**a.tho**		a.mil	7	**p.tho**	9	z.sen
		9	**m.nig**		a.mil		p.tho	10	m.nig
		10	**a.vit**		n.oli		a.mil		
				9	s.mon	8	m.nig		
				10	**b.mix**	9	**l.ful**		
							s.mon		
						10	t.liv		
	New spp	10		7		5		5	
	Cumulative spp	10		17		22		27	

Abbreviations are used for all scientific names. All four lists contain 10 different species. New species are marked in **bold** and italic. Repeat records of the same bird within one of the lists are omitted from calculation but recorded in order to allow analysis by different length lists and to better estimate relative abundance. The number of new species recorded in the lists will eventually decline and the species-accumulation curve will flatten out, so that it can be determined when the site is adequately surveyed. The data can also be used to calculate relative abundances.

Reproduced with kind permission of Lars Dinesen, Regulus Consult, Copenhagen.

and Fjeldså (1999), in Tanzania, have suggested a refinement to the data recording procedure, which will enhance the analytical possibilities after the survey is complete. Their recommended field method involves recording details of the records of all individual birds on separate sheets of paper within a note book. This offers the opportunity of assembling lists from various observers and changing the size of the lists freely. Otherwise one has to decide in advance whether a 10-, 15- or 20-species list is the most appropriate for the specific locality.

In the higher altitude forests of Bolivia, Poulsen *et al.* (1997) discovered that the habitats were so poor in species that lists of 10 species were more suitable than lists with more species. The number of species recorded in the list is related to the richness of the local avifauna, and thus for montane forests of Africa 15-species lists seem appropriate, and in SE Asian forests

20-species lists are more useful (McKinnon and Phillips 1993).

The use of all bird records collected during the listing method may allow a significant refinement of the results over that resulting from an assessment of the frequency of birds being recorded on 20-species lists. Fjeldså (1999) found that there was a good relationship between the frequency of birds recorded from 1-ha plots and the frequency of birds recorded from 'random walks'. The relationship with the frequency in 20-species lists was less useful, mainly as there is a rapid peak of 'common' species which occur on all the lists and hence do not provide useful data on which of them is more common than the other – they all appear to have the same degree of abundance.

4. Area counts

Area counts use a semi-standardised 1-ha plot (or it could be larger in open habitats) as the recording area. The observer can walk around freely within the area and use skill to locate birds that hide in more inaccessible parts of the plot. The method has been developed in tropical montane areas in Africa where it is generally not possible to set out a series of random point counts because the terrain prevents access to many of the areas selected and locating all the birds from a stationary point is generally not possible and there are often more species to be found when compared with the Palaearctic region (Koen 1988; Fjeldså and Rabøl 1995). Many surveys of remote tropical forest areas are also of rather brief duration and the need to collect reasonably standardised data with only a few days fieldwork is also an important consideration.

Areas are chosen from the point of view of being able to get to them. The plot is defined roughly by the use of obvious landmarks (such as large tree trunks), although there is latitude for adjusting the boundaries as the area is walked around to take account of areas that cannot be accessed due to steep drops, or tangles of vines, etc. In the easier situations the corner points should be 100 m apart to give the area of 1 ha in the final plot. Ideally plots should be distributed randomly, but this idea has to be fitted to the realities of the situation in the forest. No plots should be closer than 100 m apart. Plots should aim to sample sufficient variation in the area (vegetation, altitude, disturbance regimes) to provide data which will meet the objectives of the study. If the study is comparing bird assemblages between habitats, there should be sufficient numbers of plots in each habitat to provide comparable data (adequate sampling can be assessed by plotting species accumulation curves and adding plots until the curve flattens out), or a minimum of 5 plots should be recorded. A description should be made of the vegetation structure within the plot (1/10–10/10 cover of field (ground) level, herb/bush level, bush level, understorey trees, canopy and emergents) to allow a test of the representatives of the plots chosen, and to allow comparisons to be made between sites with different habitats.

All birds heard or seen within the plot during a fixed time period are recorded in a notebook. Birds seen within and at a greater than 50 m distance are recorded separately. During the first few minutes of the recording period the observer stands still and quiet, as in a standard point count. In the latter part of the count the observer moves around the plot slowly and quietly and stops at suitable vantage points to look or listen. During this period the observer also uses knowledge of the habitat requirements of different types of birds to try to find them within the plot. Because movements are slow and careful they do not interfere with collecting the data, nor do they cause an appreciable disturbance to the birds. Repeated ten-minute samples can be collected until the accumulation of new species ceases.

Data can be presented both as the cumulative number of species observed in the plots, an average number of species recorded in the various plots, or as an estimate of density within the various plots based on the species records. As this is a sampling method it does not attempt to produce a complete species list or species-density (calculated using Point Count methods) for each plot, but rather to collect data which facilitates inter-plot comparisons and inter-site comparisons. However, plotting of species accumulation curves will indicate when the bird fauna has been adequately sampled. As Africa is relatively species poor in comparison with other areas it is expected that the species accumulation curves from South America and SE Asian tropical forests would take much longer to reach a plateau, and hence this method may not be suitable in such areas, although it may still work in montane areas where the avifauna is less diverse and access is difficult.

5. Which method to choose?

Methods for assessing species richness and relative abundance are not well standardised. Indeed it is a problem in interpreting the relatively few published studies on tropical forests that the methods vary so much. The X species list method has much to commend it. Fjeldså (1999) has published a modification that deals with some of the problems. Field data were collected continuously listing all bird identifications. Notes were made on unidentified calls or sightings to allow subsequent adjustment when they were identified. The data were compounded into 20 species lists to calculate overall species richness. An advantage of collecting a complete list of all records is that it allows the possibility of subsequent analysis by lists of any length (15 or 10 might be more suitable in a less speciose area). Relative abundances estimated from frequency of occurrence on 20-species lists seemed to underestimate the abundance of the most abundant species because they may occur on most lists but can still occur only once per list. Instead Fjeldså estimated relative abundance from the relative frequency from the complete list of occurrences. Thus by appropriate field recording, it is possible to use the species list method for estimating species richness but also to use the fuller data for estimating relative abundance of species. In this particular study, plots were about 1.5 km^2 in extent.

Estimation of species richness

The total number of species is a valuable summary statistic for comparing different areas. The difficulty in species rich areas is that it takes a long time to find all the species because so many are rare. Initially the recorded number of species increases quite rapidly as fieldwork progresses. Over time the rate of finding new species declines. In ideal circumstances, there will come a period when no new species are found from more time in the field – all those that occur in the area have been found. Early on, there will be quite a lot of species that have been recorded only once. These are the rarer and harder to find species. As coverage gets more thorough, there will be multiple records of even the rarer species. Stabilisation of the total number of species recorded and decline in the number with single records give confidence that most of the species have been discovered (see Fig. 6.1).

Figure 6.1

Accumulated number of species and singletons (species recorded only once) as a function of the number of 20-species lists compiled

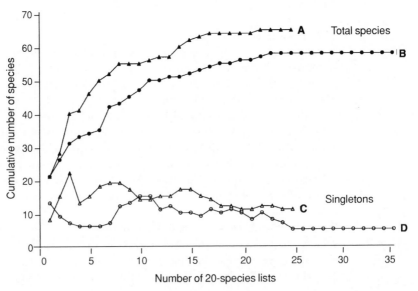

Data from primary montane forest (B) and adjacent areas with disturbed forest (A), Tanzania. Derived from Fjeldså (1999). In both habitats, the number of new species has become very small after about 25 lists which suggests near complete recording. The number of species recorded only once has reached a low level which confirms this belief for the primary forest (C). The disturbed habitats are richer in species and less fully described with several rare species that may be transients (D).

The overall species richness (SR) can be estimated as

$$SR = S_{obs} + N_1 / (2*N_2)$$

Where S_{obs} number of species observed
               N_1 number of species recorded only once
               N_2 number of species recorded twice

Further details are given by Colwell and Codington (1994).

There has been a long tradition of summarising data to a diversity index. This is not generally a helpful approach. The problem is that diversity indices do not have a clear and simple meaning. They increase with increasing species richness. They also increase with a non-even distribution of relative abundance across species. Any one diversity index may represent a different combination of these two factors. In ecological terms, species richness is the interesting parameter and is worth estimating well and without bias. Converting data into diversity indices risks clouding the meaning of results as well as hiding possible bias in the component owing to species richness.

Assumptions

1. Observers are adequately and comparably skilful

Differences in field skills among ornithologists are never more obvious than in tropical forests. A high degree of skill is needed to hear and recognise each sound and (less often) to detect birds by sight. In moderately rich areas, a minimum of one or two weeks is needed to gain adequate familiarity with the local birds. In the very richest areas, even this time would not be sufficient. Working in pairs can help to ensure that all individual birds are detected and identified and disagreements sorted out. Timed methods assume that all observers are equally quick at identifying birds. Listing methods get round this assumption to a degree by allowing the less experienced more time to make identifications. There is, however, no getting around the need for all observers to be able to detect and identify a high proportion of birds.

2. Birds are correctly identified

While correct identification is obviously necessary, it is by no means easy if field guides are inadequate or unavailable, if there are several possible related species to separate and if much identification is by voice. The most likely errors are consistent omission or inclusion of particular species by particular observers. These errors can be detected and corrected during fieldwork by careful comparison of records and discussion of inconsistencies. It is important during fieldwork to make notes about birds seen or heard but not identified. Initially they can be treated as species A, species B, etc. With luck, they will eventually come to be identified and the data can be corrected accordingly.

3. Differences in detectability have been ignored

All the methods described in this chapter ignore the fact that different species differ in their detectability. This is the main difficulty in counting birds and is the primary reason for the more sophisticated methods described in earlier chapters. Methods described here are necessary because the data are needed but the time is rarely sufficient for sophisticated studies in tropical forests. In writing and reading results, it is necessary to be aware of what can or cannot legitimately be inferred given this problem. Abundances of similar species might be compared if they are similar in detectability. Comparing abundances of dissimilar species is almost totally unwarranted. Community measures might be compared across plots. Patterns within a species might be described over time or between plots provided there is no risk of there being an important effect of habitat on detectability. Attention might be drawn to the conservation of apparently rare or missing species if, but only if, there is reason to believe that they would be detected if they were present in reasonable numbers. Otherwise, their apparent rarity may mean no more than that they are hard to detect.

4. Seasonality has been allowed for

Without extensive study, very little can be inferred about the effect of season at a particular site. Studies between sites may be biased as a result. At any one time, regularly occurring species may be absent from a site if they are migratory or they may be present but hard to detect even though at some seasons they are noisy. The only way round this is to extend the season of coverage of a study but this may be logistically impossible. In interpreting results, it is obviously important to appreciate that they strictly refer to the year and season in which they were collected.

5. Methodological details are comparable

Using different analytical methods, for instance different list-lengths, will not produce identical results even on the same data set. The best advice that can be given is to collect full data in the field as a continuous list of all birds detected. Species richness can then be estimated from appropriate length lists and individual relative abundances from all records. In interpreting the results, it is safest to remember that they very much refer to relative abundances capable of telling the common from the rare but not of discriminating at a very fine level of detail.

Examples

1. Distribution, diversity and abundances of birds in East Africa

Pomeroy and Dranzoa (1997) wanted to draw some broad conclusions about the effects of habitat factors on bird communities. They draw attention to the fact that all simple methods have weaknesses but that only big differences in species richness are likely to have practical significance especially for conservation. Their study showed that Timed Species Counts (see above) detected species at a faster rate than transect counts (Fig. 6.2).

Figure 6.2

Specie accu-
mulation
curves from
Selengi (SE)
a wooded
savannah and
Emali (EM)
a natural
grassland
devoid of
woody
vegetation in
Uganda

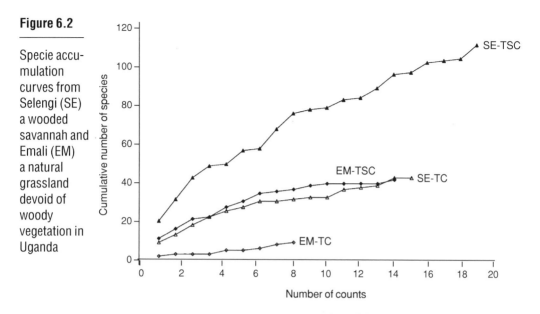

Timed Species Counts accumulate species lists faster than transects. It is clear that there are still new species being found at quite a rate and none of these surveys has fully described the bird community of the habitat. Derived from Pomeroy and Dranzoa (1997).

They argue that for many purposes, the greater rate of data collection of the simpler methods is a benefit. Both methods produced similar general findings about the effects of habitat. The longer lists produced by TSCs provided more information on the quality of individual bird communities where the occurrence of rare, threatened, endemic or characteristic species is an important consideration.

One unusual feature of this method is that species are given rankings according to the order in which they are detected within one-hour periods. There is no theoretical justification in including these weightings as the probability of seeing a bird sooner rather than later is also proportional to its abundance. This would just confound the results. If no weightings are used, the method is simply based on one-hour lists with relative abundance being determined from frequency of occurrence.

2. Impact of logging on species diversity in Sumatra

Danielsen and Heegaard (1995) wanted to assess the effects of logging and conversion to plantations on forest bird and mammal communities. They used a transect method with distance measuring but found that observers were not consistent in estimating distances especially when most (77%) records were by sound. Individual transects were 2 km in length and were walked up to 20 times, with replication both within and across days. Their unit of analysis was number of individuals per unit of effort. They drew attention to the fact that even after 40 hours, species lists were still growing and suggested that this growth would be likely to include disproportionately many rarer species of particular conservation significance. The data would probably have been sufficient to estimate total species richness in the different sites but this was not attempted.

3. Rapid assessment of Bolivian and Ecuadorian montane avifaunas

Poulsen *et al.* (1997) point out that forests are disappearing so fast that there is not time for thorough quantitative studies. They wanted to assess species richness in a variety of forests. Data were obtained by moving around along paths and open areas covering the area and altitudinal gradient as homogeneously as possible. Lists made separately by two observers were combined according to the time of records. The intended method of analysis was the 20-species list, though it turned out that in species poor habitats, 10-species lists were better. Although space consuming, field records were kept as complete sequential lists of all birds recorded. This approach makes possible the assembly of lists of different lengths and allows the combination of records from different observers.

Plots of species accumulation curves show the extent to which species lists are nearing completion and the site can be considered to be adequately surveyed. The magnitude of species richness and relative abundances of species can be estimated.

Summary and points to consider

Rigorous methods may not work fast enough in tropical forests.
Important information can be collected by simple methods.
Estimates of species richness and relative abundance may be sufficient.
Observers need to be able to identify most birds by ear.
Collecting complete data allows the best variety of subsequent analysis.
Effort needs to be standardised and reported.
Completeness of species lists need to be checked.

7

Catching and Marking

Introduction

Individual birds may be caught and marked in order to estimate population size, to investigate habitat selection and other distributions, to
calculate survival rates, measure dispersal and other movements, and to
measure the reproductive success of individual birds. This chapter concentrates on the use of catching and marking to aid the counting of birds
and the estimation of population size. The methods can be divided into
capture–recapture (in which birds are caught, marked, released, and a
proportion recaptured), and capture per unit effort (in which the effort
used to catch them is standardised, or the capture rate for a species relative
to the total number of birds captured is used for calculating population
indices). Some developments of studies involving capture, such as the use
of matrix models, are also given.

There are many difficulties and assumptions involved in catching and
marking birds. These include the need for training, legal licensing,
expertise in applying marks, and experimental design. In respect of the
latter, for instance, habitat structure, particularly vegetation height, will
influence what you catch. However, properly thought out and organised
marking experiments can provide information of immense value which
would not be produced by any other method.

This chapter gives some guidance as to when it is necessary to mark birds
and how to go about it. References to licences and permits refer to work
undertaken in Britain. Similar, but by no means identical, considerations,
apply to almost every other country. For many mobile species
colour-marking systems are controlled through international agreements.
These are set up to protect the research workers from duplicating marks
and so flawing each other's work and also to facilitate the exchange of
sightings. In some countries, including Britain, cooperation with such
protocols is mandatory for anyone obtaining a licence to mark birds, and
every research biologist considering such marking should cooperate.

Considerations before undertaking catching and marking work

A number of preconditions need to be met. Not all methods require that individuals should be recaught. In some cases field observation is adequate.

1. Will it be possible to catch enough individuals to obtain worthwhile results? Some species are far easier to catch than others.
2. Will the mark harm the bird or affect its behaviour? Will it make the bird more vulnerable to predation, alter its place in any hierarchy (e.g. feeding), or interfere with pair bonding?
3. If the bird has to be handled again for the ring to be read can it be caught again? Are the chances of capture affected by the fact that it was caught in the first place?
4. If the bird is to be observed in the field are the marks used properly distinguishable? They may fade or fall off, your co-workers (or you yourself) may be colour-blind. Is your recording system as foolproof as possible? Is the range at which you can distinguish the marks distant enough to be useful?

Figure 7.1
───────────
Marking methods

These are listed with description, constraints, advantages and disadvantages. Readers are referred to the BTO Ringers' Manual (BTO 1984) for details of field etiquette, catching methods and recording. This information has international relevance.

1. Metal ring
Each split metal ring has a return address and unique serial number. The bird is uniquely marked but must be recaught to establish its identity.

2. Colour ring
Celluloid springy rings are used on small birds but more modern materials (mostly darvic) are used for bigger birds. The most usual style is a spiral-flat band. Rings may be placed above the knee on waders to provide a greater number of colour combinations. Celluloid rings (only material available for small birds below about 30 g) may fade. Larger celluloid rings are cemented or sealed but smaller ones are not.

3. Leg flag
Coloured sticky tape around the metal ring can be used, though it falls off after a few weeks or months. This is a problem since the exact life of any tag applied cannot be known. On most types the flag sticks out about as far as the height of the ring. Cavanagh and Griffin (1993) found that Velcro tags are unsuitable for Herring and Great Black-backed Gull productivity studies but may be useful for short-term studies, as tags tend to fall off after a time.

4. Patagial tag
A small plastic flap is pinned on the top surface of the wing by a stainless steel pin through the bird's patagium and held in place with a nylon washer. Such tags may be distinguished by their colour, letters, numbers or symbols stuck or drawn on their surface. For many large species such as dabbling ducks they have been shown to be safe but are not recommended for diving birds and have never been permitted on anything smaller than a Starling in Britain. They may not be easily visible as many birds preen them into the wing coverts. Particularly bright colours used on patagial tags may make the marked birds significantly more vulnerable to avian predation, though evidence is sparse. Maddock and Geering (1994) showed that patagial tagging of nestling egrets was found not to affect fledging ability of nestlings nor did

tagging increase nestling mortality in the colony. Adult mortality was, however, increased. Patagial tagged egrets also undertook long-distance migration similar to untagged birds.

5. Neck collar

A large colour ring can be put around the neck of the bird. These have been supposedly successfully used on geese and swans but have also been the subject of significant problems, e.g. affecting pair bonding or causing physical distress and choking. Martella and Navarro (1992) describe the marking of Greater Rheas in Argentina using neck collars. Birds were 'dazzled' at night with a light beam and fitted with a numbered adjustable cattle-type legband. They were caught with a traditional 'boleadoras' which consists of either 2 or 3 balls of rounded soapstone lined with raw leather and fastened firmly to a long strap of braided leather, tied together at their other end. These were thrown at the Rheas' legs, winding round them in the process. Weiss et al. (1991) found over 90% of Canada geese marked with neck collars were re-sighted in error, which constituted 3% of the re-sightings. These errors significantly increased estimated numbers of marked geese in the sample, though estimates of survival rates were not affected by these errors.

6. Plumage dyes

Feathers do not dye easily. The best dye is picric acid which dyes light-coloured feathers orange-yellow. The marks are temporary (until the next moult at the latest). Belant and Seamans (1993) describe the technique of applying colour marks to the eggs of Herring Gulls in order to dye their feathers and aid their study. An oil-based silica gel is the recommended dye carrier but it must be applied to only one egg late in incubation in order to reduce mortality of embryos, or an artificial egg can be used.

7. Radio-tracking

This is an unrivalled technique for locating individual birds to determine home-range, time budgets, habitat selection, etc. (see Chapter 10). This equipment is commercially available, and further details of considerations in a radio-tracking study are given later in this chapter. It has the disadvantage that only a few individuals can be tracked at a time by one observer, unless expensive automated tracking is used.

8. Individual marks

Some species of birds have individually identifiable marks, which makes recognition possible without the aid of an artificial marker. The bill patterns of Bewick's Swans are a classic example (Fig. 7.2).

9. Others

Bill plaques, back tags, imped feathers, numbers painted on the faces of swans or the shields of Coot have all been used. These rather specialist and individual marking methods are not described further in this book.

Capture–recapture methods

The two main reasons for catching and marking individuals within a population are (1) to estimate population size and/or define migration routes and (2) to estimate survival rate. Analytical methods for mark–recapture data for each of these have been developed and the underlying principles, largely concentrating on population estimation which is the focus of this book, are outlined here.

Estimation of population size is based on the assumption that, if a proportion of the population is marked in some way, when it returns to the original population complete mixing occurs. A second sample is then taken. The number of marked individuals in the second sample should then have the same ratio to the total numbers in the second sample as the

Figure 7.2

Naturally
'marked'
birds

| 'Yellowneb' | 'Pennyface' | 'Darky' |
| Bewick's | Bewick's | Bewick's |

Sir Peter Scott, of the Wildfowl and Wetlands Trust, showed that bill patterns
on Bewick's Swans can be identified by variations in their black and yellow bill
markings, which usually fall within the three major categories illustrated (from
Scott 1981). Reproduced with kind permission of Lady P. Scott.

total number of marked individuals originally released has to the total population. As the number originally caught, the number marked and the number of marked individuals in the second sample are known from catching and marking, an estimate of the size of the total population can easily be calculated. Mark–recapture population estimation is particularly useful where absolute estimates of population size are needed for species that are difficult to count in the field.

Ornithologists have not made much use of mark–recapture methods (e.g. Jolly 1965) for estimating population size (largely because the models seem complicated, birds migrate or frequently appear to violate assumptions of the methods, and a lot of effort is required for small returns), although reviews can be found in Cormack (1968, 1979), Seber (1973) and Nichols *et al.* (1981). Mark–recapture models can be classified according to assumptions about whether the population is closed, i.e. the population is not influenced by mortality, recruitment or migration (both immigration and emigration). 'Models' or mathematical expressions are used to define two broad classes – for closed and open populations. Models such as the Lincoln index and Jolly–Seber are used to analyse capture–recapture data and this section concentrates on their applicability and assumptions rather than mathematical detail, which is referenced later in the chapter.

There are four classes of mark–recapture models:

1. Closed populations:
   (a) two-sample experiment (e.g. Lincoln index type model with two capture events)
   (b) *K*-sample model (many capture events)
2. Open populations:
   (c) completely open populations (both losses and gains, e.g. Jolly–Seber model)
   (d) partially open populations (most commonly losses and no gains, but also gains and no losses)

To satisfy the requirements of the various methods of analysing mark–recapture data for population estimation, and having defined why it is to be performed, a number of assumptions of the methods should be considered. The identifying number of each assumption is related in Table 7.1 to the four types of model described above. The ways in which these assumptions are likely to be violated and possible ways of reducing the problems are presented below.

Table 7.1

Assumptions of the four classes (a–d) of mark–recapture models

Assumption	a	b	c	d
1. Closed population	X	X		
2. Equal capture probability	X	X		
3. Marking has no effect on catchability	X	X		
4. Second (subsequent) sample(s) are random	X			
5. Marks are permanent	X		X	X
6. All those marked that occur in second sample are reported	X	X	X	X
7. Capture probability assumed constant for all time periods		X		
8. Same probability of being recaught in all capture events			X	X
9. Equal probability of survival			X	X
10. Equal probability of caught birds being returned			X	X
11. Sampling time is negligible			X	X
12. Losses from emigration and death are permanent			X	
13. Population closed to recruitment only				X

For example, models a and b both assume that the population under investigation is 'closed', i.e. there are no additions or subtractions (immigration/births or emigrations/deaths), and so on.

Note: a = Two sample model – closed populations.
     b = K-sample model – closed populations.
     c = Completely open populations.
     d = Partially open populations.

Assumptions

1. The population is either closed or immigration and emigration can be measured or calculated

A closed population is one where there is no immigration or emigration of birds during the period of population estimation. There should also be no births or deaths within the sampling period unless allowances can be made for them.

Likely causes of violation: immigration, emigration, births or deaths occurring at the time of the study may be exacerbated by captures being taken over a long period.

Reducing the problem: reduce time intervals over which captures are made and conduct captures at a time of the year when migration and recruitment of young birds to the population are not occurring. Sampling should be at discrete time intervals and the time involved in taking the samples should be small in relation to the total time.

2. There is equal probability of capture in the first capture event

All individuals of the different age groups of both sexes should be sampled in proportion to that in which they occur in the habitat. Therefore, all individuals of the different age groups should be equally available for capture irrespective of their position in the habitat.

Likely causes of violation: there may be part of the population that is never captured because the individuals concerned are trap-shy or they cannot be sampled in certain habitats. Alternatively some birds may be trap-happy, i.e. caught very often.

Reducing the problem: if it is possible that the probability of capture differs between the sexes or other 'sub-groups', population estimates should be derived independently. For example, female pheasants are trap-happy and more likely to be captured than males while living in harems prior to egg-laying. The regular movement of traps within the study area reduces the over-sampling of these individuals. It may be desirable to use traps to catch the first sample and to apply highly visible markings so that the second sample can be obtained by re-sighting rather than capture. This has been used successfully on Mallard (United States), Pheasants (United Kingdom) and Willow Ptarmigan (Red Grouse) (Scandinavia).

3. The marked birds should not be affected by being marked

Likely causes of violation: marks significantly affect behaviour, e.g. neck collars have been reported to contribute to starvation in Snow Geese. Higher re-sighting probabilities have been suspected for patagial-tagged, back-tagged and dyed birds, than for unmarked birds, in studies in the United States. Increased predation of birds tagged using patagial markers has also been recorded.

Reducing the problem: be aware of these effects and remedy by changing the marking system.

4. The population should be sampled randomly in subsequent capture events

Likely cause of violation: individuals do not mix randomly and perhaps change the area they use, biasing future capture.

Reducing the problem: this is a difficult problem to overcome. More traps and more field observations are desirable.

5. The marks should be permanent

Likely causes of violation: marks fall off or become unreadable at a distance.

Reducing the problem: be aware of unsuitable designs which have a high failure rate. Double mark where necessary (e.g. a leg ring or band as well as a back-tag). On future capture worn or lost markers can be replaced.

6. All marked individuals occurring in the second or subsequent samples are reported

Likely cause of violation: this assumption is generally relevant to experiments in which the second sample is based on ring or band recoveries made by the general public (e.g. hunting recoveries).

Reducing the problem: the total number of recoveries for use, for example, in Lincoln-index model estimates requires an assessment of the reporting rate, i.e. the proportion of recovered rings or bands that is reported. Reporting rate has been estimated using either additional information on the number of recovered rings obtained from hunter questionnaire surveys, or reward studies in which some rings or bands are marked with a message that a reward is offered for their return.

7. Capture probabilities are assumed constant for all periods

Likely cause of violation: capture probabilities will vary from one capture event or sampling period to another, as a result of weather factors and possible changes in sampling effort.

Reducing the problem: conduct sampling under the same weather conditions wherever possible, and use the same sampling effort, e.g. length of capture net, number of traps, etc.

8. Every bird in the population has the same probability of being caught in sample i

This assumes that the bird is alive and in the population during sampling period i.

Likely causes of violation: age-specific and sex-specific variation in capture or habitat use.

Reducing the problem: sample the different age, sex and other 'sub-groups' independently.

9. Every marked bird in the population has the same probability of surviving from sampling periods i to $i + 1$

This again assumes that the bird is alive and in the population immediately after the time of release in sample i.

Likely causes of violation: age-specific and sex-specific variation in survival, e.g. higher mortality of female waterfowl (as shown in studies in the USA for example), caused by differential predation.

Reducing the problem: sample the different age, sex and other 'sub-groups' independently.

10. Every bird caught in sample *i* has the same probability of being returned to the population

Likely causes of violation: differential age, sex and individual stresses causing mortality during handling.

Reducing the problem: be aware of experiences and keep handling time to a minimum.

11. All samples are taken instantaneously such that sampling time is negligible

Likely causes of violation: this assumption is never strictly met.

Reducing the problem: as previously, keep the sampling period short.

12. Losses to the population from emigration and death are permanent

Likely causes of violation: birds that have emigrated return to their natal site, showing high site fidelity (common in birds).

Reducing the problem: use capture and re-sighting efforts. Radio-telemetry is a useful technique in distinguishing between dispersal movements and emigration, and direct mortality.

13. Population closed to recruitment only

Likely causes of violation: recruitment occurs but goes unnoticed.

Reduction of problem: choose the time of year when recruitment does not occur. This has been successfully achieved in American Woodcock, using a model known as the 'death but no immigration model'.

Estimating population size

There are many methods for estimating population size, and only a few are given here. They are (1) the simple Lincoln index, (2) the du Feu method, and (3) other methods in which there are more than two sampling occasions.

1. The simple Lincoln index

The simplest Lincoln index model is based on one capture and one recapture only, i.e. a two-sample case, although computations for a multi-sample case in which a number of sub-samples are taken are described below. Variants of the method have been derived to allow for losses (emigration and death) or gains (immigration and birth) to the population. Provided the conditions listed under model a in Table 7.1 are satisfied, the total population, size P, can be calculated from the simple Lincoln index as shown in Fig. 7.3. The method is most appropriate for

Figure 7.3

Calculation
of the
simple
Lincoln
index

The simple Lincoln index is used here to analyse a 'two-sample' case of capture–recapture data to provide an estimate of population size, P. The calculation of standard errors is also shown. The general format is: Number in population/Original number marked = Number in second sample/Number recaptured, or

$$P = \frac{a\,n}{r}$$

where n = number of individuals in the second sample, a = number marked and r = number recaptured.

If the second sample (n) consists of a series of sub-samples and a large proportion of the population have been marked, recovery ratios (= r/n) can be used to calculate the standard error (from the variance) of the estimated population where:

$$P = \frac{a}{R_T}$$

where R_T = the recovery ratio (r/n) based on the number of birds in all of the samples; and the variance is approximately:

$$\text{var } \hat{P} = \left(\frac{a}{R_T{}^2}\right)^2 \times R_T \; \frac{(1 - R_T)}{y}$$

where y = number of birds in the sub-samples. The standard error of the estimate is the square root of the variance.

The method is based on **direct sampling** in which the number in the second sample is predetermined. Alternatively, **inverse sampling,** where the number of marked birds to be recaptured is predetermined, has the advantage of giving an unbiased population estimate and variance in cases where the number of recaptures is small:

$$P = \frac{n\,(a + 1)}{r} - 1$$

An approximate estimate of the variance of this is given by:

$$\text{var } \hat{P} = \frac{(a - r + 1)\,(a + 1)\,n\,(n - r)}{r^2\,(r + 1)}$$

Further details of this group of methods are given in Southwood (1978).

Example of estimating population size in the Pheasant using the simple Lincoln index

Consider a simple example in which, during the winter of 1984/85, 27 male and 70 female Pheasants were marked using plastic back-mounted fin-tabs, each bearing an identifying number. Trapping was conducted continuously each day and as more birds were marked, the number of recaptures increased. The traps were not moved and were baited with grain to attract birds into them. Although the primary aim was not to estimate population size from these data, but to investigate behavioural trends in spring, it is useful to consider the application of the Lincoln index method in estimating population size. A comparison will then be made with that for the spring population using the marked birds observed from a vehicle in daily morning and evening watches.

Catches made after the 70 females were marked included 62 marked birds out of 66 birds caught.

$$P = \frac{70 \times 66}{62}$$

= 74 females = size of winter population.

Catches made after the 27 males were marked included 21 marked birds out of 24 caught.

$$P = \frac{27 \times 24}{21}$$

= 31 males = size of winter population.

This gives a total winter population of 105 birds.

The main limitations of this method are assumptions 1–6 in Table 7.1. Those of population closure, no mortality and equal capture probability, are most important. Pheasants are relatively sedentary during the winter although immigration and emigration occur during late winter and early spring (and in some cases throughout the winter). Further, the period over which capture took place was too long and some mortality due to Fox predation was known to have occurred. Some individuals, particularly hens from the same harems, were trap-happy and usually a greater proportion of hens from the same harem were caught than would be expected by chance, since they feed together during the winter and early spring period. For these reasons the use of the Lincoln index is not recommended in this particular case.

The spring censuses of Pheasants on the study area were used to estimate a spring population size for, approximately, a 1-month period by mapping all birds seen on successive morning and evening visits throughout mid April–mid May. The position of marked and unmarked males and females (together with identifying code for those marked) were entered on a map. The proportion of males and females on the census that were observed tabbed (CT) for the whole of the spring observation period are shown below. The total number (TT) of males and females tabbed on the study area during the catching period is also shown. Simple Lincoln index calculations permit the estimation of population size of the sexes separately. In this case recapture events are actually resighting events.

The Lincoln index method used on resighting data for Pheasants rather than on recapture data:

Census birds tabbed (CT as percentages)	50 males, 59 females
Total birds tabbed (TT)	27 males, 70 females
Estimated population $\left(\frac{100}{CT} \times TT \right)$	54 males, 119 females

From this it can be suggested (with the proviso of population closure, no mortality and equal capture probability) that both the female and male 'populations' have been added to from late winter to early spring. This is one of the many benefits of having birds marked with visible tags which allows 'sampling' to be observational without capture.

To summarise the Pheasant example, there are various assumptions with this method, most important of which are (1) that the marked birds must freely mix within the extant population; (2) all birds should have an equal probability of being seen; (3) tags must be permanent and must not fall off; and (4) no mortality should occur from the period of capture and marking to the period of resighting. These correspond to assumptions 4, 2, 5 and 1, respectively, in Table 7.1.

studies of colonial species which cannot be counted directly such as a seabird colony (e.g. Storm Petrel) or warblers in a reedbed. In both cases the boundary of the habitat is known and can be defined. It is perhaps least useful for populations that are continually changing in number, or for birds in more complex habitats such as gardens.

2. The du Feu method

The standard methods of mark–recapture analysis, such as Jolly's (1965), require repeated samples being taken at fixed intervals from which measures of immigration and emigration may be estimated. The du Feu method (du Feu *et al.* 1983) can be used to estimate the population of a species, such as a warbler in a wood, being ringed during a single session e.g. day, week or season, depending on the mobility of the species concerned. If the species is sedentary, the session can be a winter or other field-season measure. Details of the method are given in Figs 7.4 and 7.5.

Plots of typical running du Feu estimates are shown in Fig. 7.6 for four scenarios: (a) for a closed population wide fluctuations are rapidly damped to a relatively stable value; (b) for declining populations the estimate slowly

Figure 7.4

The du Feu method of analysing capture–recapture data

Du Feu et al. (1983) present a method for making 'spot estimates' of populations in which the number of new birds (N) and the number of recapture events (R), over S captures, are used. The number of new birds (N) is the number of birds captured for the first time in the session. If a bird is captured four times in a session it contributes 1 to N and 3 to R. The equation is:

$$1 - \frac{N}{P} = \left(1 - \frac{1}{P}\right)^{S}$$

where S is also equal to N + R. The equation has no explicit solution so P has to be found iteratively.

One begins by making a sensible guess. Then, with successive sessions, and by inputting N and P from them, one can use a program to make a running estimate of the population. A graph of running estimate (P) against capture event (N + R) will show if the estimate is stable, which can indicate when the trapping should be stopped, since more captures would not alter the estimate. Such a plot would also indicate whether the population is closed in that, for example, immigration would give rise to a decrease in recapture events and an increase in new birds. An example of how the estimates are refined as the number of recaptures increases is shown in Fig. 7.5.

The standard error of P is:

$$SE_P = \sqrt{\frac{P}{e^{\left(\frac{N+B}{P}\right)} - 1 - \frac{N+R}{P}}}$$

Example of the use of the du Feu method on the Pheasant data

The values of N (the number of new birds caught) and R (the recapture events) for the same session i.e. winter period, gave du Feu population estimates of 51 ± 12 for males and 89 ± 6.6 for females, with a total population estimate of 132 ± 9.7 birds. Running estimates using the du Feu method could also be made for successive capture events through the catching period. The Lincoln index method gave a total population estimate of 105 birds, compared to 132 birds using the du Feu method.

Figure 7.5

The du Feu
method for
calculating
population
size

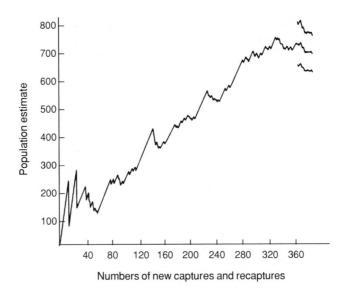

In this method, each bird captured, whether it is a new bird or a recapture, enables a new population estimate to be made. If each successive estimate is plotted against the capture event it is possible to follow fluctuations in the estimate, and to identify the point at which it settles down. Confidence limits are also calculated and plotted (from du Feu et al. 1983).

Figure 7.6

Examples of
du Feu
estimates of
population
size against
number of
capture
events

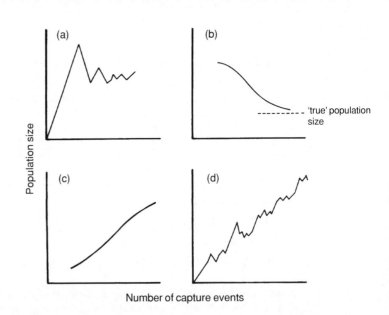

(a) Population closed; (b) population with emigration; (c) population with steady immigration; (d) population with step-wise immigration (from du Feu et al. 1983).

declines, but is always greater than the true population – the estimate can never be less than the number caught (N), although in a rapidly declining population there may be fewer birds present at the end of the season than have been ringed; (c) for smoothly increasing populations the estimate increases smoothly; (d) for a stepwise increasing population the estimates consist of a series of steps.

The conditions and limitations of the method refer to assumptions under class b in Table 7.1. In particular the method assumes population closure, equal probability of capture, no modification to behaviour caused by trapping, and additionally (assumption 11 of Table 7.1), negligible handling time of the birds relative to the study period. In respect of the assumption of population closure, trapping should be carried out for a long enough period to obtain a reasonable number of recaptures, but not so long as to risk immigration and emigration. In isolated small habitats it is possible that the whole population is being sampled, whereas in large, uniform habitats the estimated population may be an accurate measure of the catchment area population, but that area may not be known. Where individuals are being caught at an exploited resource, e.g. a water source, it is the usage that is being estimated, not the population.

Underhill and Fraser (1989) have developed a Bayesian analogue of the du Feu estimate and have used the method to estimate the number of Malachite Sunbirds at a flower food source in South Africa. The assumptions are identical to those of the du Feu method but the Underhill and Fraser method is computationally simpler.

Bayesian methods assume some prior knowledge about the maximum size of the population being trapped whereas the du Feu method does not require prior knowledge or assumptions of maximum population size. As new birds are caught the method calculates the new probabilities of there being x birds in the population. Each unringed bird captured shifts the probability distribution to the right, increasing the estimated population size; each retrapped bird shifts the probability distribution to the left, leading to a decline in the estimated population size.

The method has a number of advantages over that of du Feu, and is likely to be used more extensively in the future, in preference to the du Feu method. First, it provides more realistic confidence limits because they are based on an exact probability distribution. Second, the method is better than that of du Feu for small populations (<100 birds). The main disadvantage of the method is that if the initial guess of the maximum population is too low, the population estimate will approach its asymptote at the far right of the plot, therefore it is important not to make a serious underestimate from the outset. There is no harm done to the estimate of population size if the maximum population is overestimated, however.

3. Other methods with more than two sampling occasions

For useful reviews see Nichols *et al.* (1981) and Pollock (1981). These methods are generally multi-sample Lincoln index type models and the sampling regime is similar to a Lincoln index two-sample case in that birds are captured during an initial sampling period, marked and returned to the population. A second sample is then taken (e.g. on the following day) and recaptures of marked birds are noted. New captures are also given marks and all birds are returned to the population. This procedure is repeated for *K* sampling periods. The main difference between these methods and the Lincoln index two-sample method is that each bird must be given an individual mark, e.g. serially numbered leg ring. The models used to describe recapture data from *K* sample studies require complete capture histories. The probability distribution for the set of possible capture histories is then expressed using a multinomial model treating population size and capture probabilities as parameters. Examples of models for closed populations are described in Otis *et al.* (1978). Assumptions of these types of models relate to class b in Table 7.1. For short-term studies, closed population models allow for unequal catchability of individuals (see Pollock 1981).

Models that take account of population gains and/or losses, i.e. for completely open populations have been developed by Jolly (1965) and Seber (1965). These are denoted the Jolly–Seber stochastic models for completely open populations. Both population size and survival rates are calculated. The model has not been extensively used because of violation of a number of the assumptions presented under class c in Table 7.1. Further, a better survival model has been developed by Clobert *et al.* (1987), although population size estimation is not a feature of their model. For long-term studies, open population models that assume equal catchability are used (see Pollock 1981). Such models allow estimation of survival and birth rates as well as population sizes.

Mark–recapture models for populations that are open to both gains and losses, as described in the last paragraph, have the greatest potential applicability to studies on bird population dynamics. Models do, however, exist for populations that experience only losses, and no gains, and which are subject to the assumptions in class d of Table 7.1. See Nichols *et al.* (1981) for further descriptions.

Methods based on catch per unit effort

Methods based on capture per unit effort rely on standardising effort of capture or observation. The aim is to estimate a species abundance by making the effort by which the abundance data are obtained constant. The main assumption is that the standardised design yields unbiased data from which to calibrate population size or an index thereof. Seber (1973) provides a review of the method for closed and open populations.

1. Where ringing effort is standardised
By controlling, i.e. standardising the effort invested in catching and marking, detailed studies of (1) population size, (2) productivity and (3) survival can be made. The Constant Effort Sites Scheme (CES) of the BTO aims to collect data to investigate changes in these variables by using a series of constant net-sites worked on each of 12 standard visits spread between May and August. The following are investigated.

1. Index of population change – changes between years in the numbers of adults captured.
2. Productivity – the ratio of juveniles to adults captured late in the breeding season.
3. Survival – between-year retraps of ringed birds.

The same ringing and netting sites, with the same length and type of net, are used each year. No other netting is carried out within 400 m of the net-sites. Where possible, sites within a single major habitat are preferred. Continuation of the study is important in order to minimise fluctuations in population indices resulting from changes in the habitat or geographical composition of the sample, and to allow comprehensive survival estimates to be calculated.

Netting is conducted for a set time of about 6 hours per visit and no tape lures or baits are used (as in other studies) to attract birds to the site. Habitat is recorded on 1:2500 scale maps. The validity of measuring population changes by mist-netting is assessed by comparing results with those from point counts (see Chapter 5), which are conducted at the site in the time between visiting individual nets.

The results are interpreted by (1) calculating changes in the number of adults caught between years using the same sites, to provide a measure of change of the size of the adult population; (2) calculating the proportion of young birds caught on the same sites as an estimate of productivity in the post-fledging period; (3) calculating survival rates using the SURGE routines described in Clobert *et al.* (1987).

Fig. 7.7 shows an example of a CES population index in which the base year of 1986 is taken as 100. Fig. 7.8 shows an example of productivity monitoring from CES sites in which the proportion of those caught represented

Figure 7.7

Population indices derived from the UK Constant Effort Sites Scheme

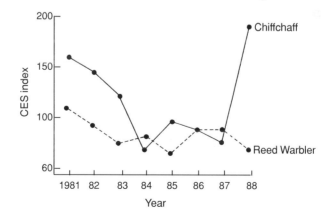

The base year of the index is 1986, and Chiffchaff and Reed Warbler indices are shown from 1981 to 1988 (from Peach and Baillie 1989). Reed Warbler populations have remained stable and Chiffchaff populations have increased, for unknown reasons.

Figure 7.8

Productivity
monitoring
using the
BTO
Constant
Effort Sites
Scheme

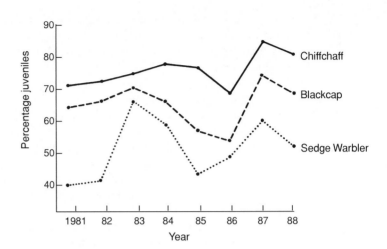

Changes in the percentage of juveniles caught under constant trapping effort at specific sites are shown for Chiffchaff, Blackcap and Sedge Warbler for 1981–1988 (from Peach and Baillie 1989). These show broadly similar trends.

by juveniles is plotted each year. Significant changes in the population index and proportion of juveniles can be calculated. One of the weak points of the CES scheme is that the sites chosen may not be representative of the population as a whole of the species being studied. CES should pick up general trends across sites but may not be good at detecting within-site differences.

Few studies have looked at whether more birds are caught in mist nets in years when there are lots of nestlings (more nestlings ringed), which is an important assumption if CES ringing is to provide useful information on a population's productivity both at a site and nationally. du Feu and McMeeking (1991) investigated whether constant effort ringing provided a good estimate of juvenile abundance. They found a significant correlation between the numbers of young birds caught in mist nets and the numbers of nestlings fledged, though there was significant variation between species. This study indicates that, within reason, CES results can be used for productivity monitoring. Peach *et al.* (1990) analysed long runs of data on Reed and Sedge Warblers from two ringing sites. At both sites more individual Reed Warblers were retrapped in total in years after ringing than there were total recoveries of dead birds from the entire British ringing scheme. Hence analysis of retrap data yields far more information than relying on dead bird recoveries. They showed that standardisation of catching effort leads to substantial improvements in the precision attainable on survival estimates.

Another example of a programme in which ringing effort and habitat is standardised is that of the 'Mettnau-Reit-Illmitz' (MRI) scheme (Berthold *et al.* 1986). The three sites are in central Europe (S and N Germany and E Austria) and trapping of passage migrants has taken place from the end of

June to the beginning of November since 1974. Trapping takes place daily, for the same length of time per day, using the same netting system and the same length of nets at the same number of net-sites. Vegetation is trimmed to maintain consistent profiles between years so as to halt the effects of successional changes in habitat. Busse (1990) notes that habitat bias can, however, extend further out from the immediate site as an area around the ringing station may strongly alter its attraction to birds, though this bias cannot be calculated. Five regression models were used in the MRI scheme to analyse correlation coefficients for regressions of total numbers of individual species captured against year, 64% of the coefficients being negative implying population declines.

Fig. 7.9 shows a diagram of two hypothetical netting operations within scrub habitat. At one site the same net length is used, visits are made every week, for the same length of time every day, in similar weather, and vegetation profiles are trimmed to keep the habitat at the same successional stage. The other site uses variable net length, different observers, variable work day length and variable number of visits through the season, and the habitat is succeeding to woodland. Population indices for Whitethroat calculated from captures from the two sites are not very similar. Which one should we trust? The data collected under standard effort is more trustworthy, but still may be affected by habitat changes in the nearby vicinity.

Ormerod *et al.* (1988) used a constant length of mist-net and constant catching effort to sample Dipper, Kingfisher and Grey Wagtail along river systems in Wales. The effective width of the river sampled was also recorded. The results for Dippers were calibrated with the known abundance. Mean catch per hour was correlated with the number of birds per 10 km stretch ($r = 0.99$), as was the mean catch per hour standardised per 10 m net ($r = 0.96$). The study concluded that standardised ringing along rivers can be used in population monitoring and suggested that the method may be effective in assessing annual change in breeding success, post-fledging survival and overwinter survival.

Kaiser and Bauer (1994) undertook a comparison of four different methods – point counts, line transects, territory mapping and capture–recapture, to determine the breeding population of two study plots in Germany. Capture–recapture with mist nets was well suited to providing good estimates of population size of breeding passerines and, being especially suited to secretive species most active within a few metres above ground, has clear advantages over visual/acoustical censusing.

2. Where ringing effort is not known

Annual ringing totals can be used to analyse population trends. The main assumption is that there is a relationship between changes in annual totals ringed and the abundance of a species when corrections are made for ringing effort. Fig. 7.10 shows how to correct annual variations in ringing effort for individual species, and therefore how to relate this to variations in a species' abundance.

A similar approach has been used for data for the Stock Dove in Britain (O'Connor and Mead 1984). The number of nestling Stock Doves ringed

Figure 7.9

Importance
of standard-
isation in
netting
operations

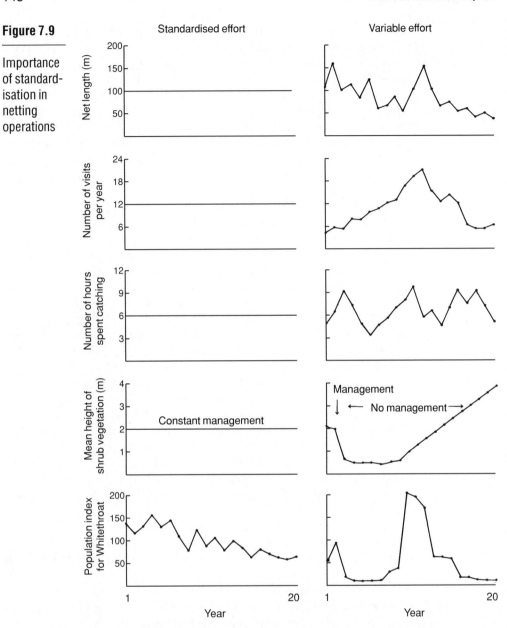

The diagram is of two hypothetical netting operations within scrub habitat – one
where capture effort has been standardised and the other where capture effort
has been variable over the past 20 years. Population indices for Whitethroat
calculated from captures are different, yet the indices for Whitethroat in the
standardised-effort system were used to calculate those in the variable-effort
system. The standardised-effort system shows a decline in numbers, the
variable-effort system does not. We should put more trust in the index trend for
the standardised-effort system whereas nothing useful could be said about
population trends from the results from the variable-effort system.

Figure 7.10
_____

Capture per
unit effort
where
ringing
effort is not
known –
correcting
for annual
variations in
ringing
effort

To correct for annual variations in ringing effort, the numbers of individual species ringed each year can be transformed into numbers of each species ringed per 100 000 birds ringed, referred to as the standardised annual totals according to the formula:

$$Y'_n = \frac{Y_{n-2} + 2Y_{n-1} + 4Y_n + 2Y_{n+1} + Y_{n+2}}{10}$$

where Y_n is the annual total of birds ringed in year n, and Y'_n is the weighted running mean of year n.

Variation in ringing effort from year to year is difficult to estimate since few ringers or observers record the total length of net they use, the total number of hours they spend ringing, or the variation in weather during ringing sessions. To correct for this variation the total number of birds of all species per year is used, and individual species totals are adjusted accordingly to remove any increasing or decreasing trend in totals ringed which, it is assumed, is due not to increases in bird populations but to an increase in ringing effort. An example of the use of this method to identify population trends in birds ringed in Sweden is given in Österlöf and Stolt (1982).

Figure 7.11
_____

Comparison
of popula-
tion indices
produced by
different
methods

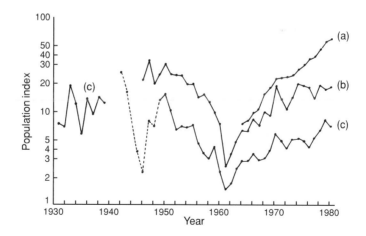

This figure presents the results of three methods of population indexing for the Stock Dove in Britain since 1930.

(a) From Common Birds Census; (b) from ringing totals; (c) from nest record card totals (from O'Connor and Mead 1984). Dashed lines are periods with limited nest card returns.

annually per 1000 nestlings of all species ringed nationally during 1931–80 was calculated. The all-species totals alter with ringing effort, but, with a large number of species involved they are likely to average out species-specific fluctuations. In this analysis information was further compared to annual ratios of nest record cards submitted to the BTO (dating back to 1930), as well as the Common Birds Census Index (dating back to 1962). Fig. 7.11 shows the agreement between the three methods of population indexing since the early 1960s, and can usefully suggest population trends prior to censusing (i.e. pre-1962).

Mist netting for censusing forest birds

A number of studies have used mist netting to census birds in lowland or higher altitude tropical forests (Bierregaard and Lovejoy 1988, 1989; Terborgh *et al.* 1990; Poulsen 1994; Remsen and Good 1996). Some of the methods are described in Chapter 6. In the latter example, long-term data obtained during a study of mixed-species bird flocks were used to evaluate the advantages and disadvantages of mist netting to describe species richness and abundance, in a bird community which is difficult to census.

In this study, nets were placed in continuous rows in primary forest, but in front of 5–6 m tall secondary vegetation along semi-open transects such as trails, between 0600 and 1800 hours, over four 5-day periods. Effort was calculated in net-metre-hours (NMH). Biases included fewer captures with increased daily hours of rain and wind (though cloud forest birds in mixed species flocks are more active during rainy days); time of day, with more caught in the morning; recapture susceptibility; and sampling bias (as a result of a monotonically damped increase in the number of species recaptured with an increase in total number of captures, termed the species encounter function).

One advantage of mist netting is that the obvious biases associated with other censusing techniques (visual and acoustic) are avoided. Mist-net sampling also enables quantitative comparisons of secretive or rarely vocal species, or non-territorial species that are inappropriate for some census techniques.

Disadvantages are that only some 45% of the species in the above study area were captured in nets – those not captured included swallows, swifts, nocturnal species, raptors, parrots and other large species. Some 25% of 64 species mist netted were not recorded by other methods (observations and voice identification). Even with substantial effort, only about a third of the avifauna was recorded by mist netting. Consequently, mist netting produces a highly biased estimate of bird species richness with canopy species being particularly under-recorded. Capture rates can be compared only if the time-effort is the same since birds learn to avoid nets, leading to a decline in the number of new captures after the first capture-day. Higher relative capture rates therefore occur when capture effort occurs within short periods of time. A further common problem with mist-netting studies is that relative capture of birds is typically equated directly with relative abundance (Remsen and Good 1996), which is incorrect. Some of the reasons for this flaw are that (a) ground level mist nets sample only that portion of the avifauna that moves within 2–3 m of the ground; (b) territorial and wandering species have different probabilities of recapture; (c) there are differences between birds in their flight distances ('home range' size) and flight frequency.

Further developments of studies involving capture

There are two more specialised procedures which involve the use of catching and marking techniques but which fall outside capture–recapture and capture per unit effort categories. These are (1) the use of matrix models in which productivity, population sizes and survival estimates are used to predict future numbers of a species (i.e. using some of the techniques previously described), and (2) the use of radio-telemetry in the special case of studying the change in location of an individual bird and its use in calibrating more extensive but more cheaply obtained data.

1. Modelling with birth and survival rates – the matrix model
Survival and fecundity (leading to productivity) are often age-specific in certain species. The descriptions given below present one method of modelling data obtained from ringing recoveries when fecundity of age cohorts is known. The matrix model was developed by P.H. Leslie in 1945.

There are three requirements for using the matrix model: (1) the number of individuals of each age category in the population being studied (P); (2) the age-specific fecundity (B); and (3) the age-specific survival rates (S), derived, for example, with the use of the SURGE program (Clobert *et al.* 1987).

2. Radio-telemetry
This section is meant only to give an overview of the major considerations to be addressed when embarking on a radio-tracking study. For a fuller description of the technique and analytical routines see Kenward (1987). The radio-tracking system usually consists of the transmitter, which is fixed to the bird, a multi-channel receiver, which detects the signal emitted by the transmitter, and one or two hand-held portable directional or fixed antennae. Table 7.2 lists the main considerations and provides additional comment. Fig. 7.12 shows the value of radio-tracking in being able to pick up movements of the bird to new areas: observations of the tagged Pheasant from a vehicle were more restricted and did not pick up the bird within woodland or when it moved some distance to feed in a game crop. The miniaturisation of the technique means that it can now be applied even to following individual pheasant chicks (Fig. 7.13). The use of radio-telemetry in studying bird distribution at the habitat-scale, and habitat selection, is described in Chapter 10.

Table 7.2

Major considerations for undertaking a radio-tracking study

Consideration	Comment
Activity of the bird	Will breeding or feeding behaviour damage the transmitter?
Weight	How big is the bird relative to its probable range in the habitat?
Weight of transmitter in relation to body size	Should be less than a notional 5% of the bird's body weight
Harness design	Will this affect its behaviour? Tail mounted, back mounted, collar mounted, leg mounted?
Range	What is the habitat; will this reduce or improve expected range of transmitter?
How many birds to mark	Logistics of observer moving between radio-location points. How many birds can be tracked comfortably at any one time?
Special transmitters	To monitor bird activity e.g. flying or resting, upending by ducks, depth measurements in diving birds, physiological measurements e.g. heart rate
How much sampling	How many radio-locations and over what period? e.g. 30 locations over 10 days for Pheasants will define their home range adequately for analysis during this period
How to obtain data in the field	Triangulation from a vehicle, on foot; use of a data logger linked to a computer database;automatic triangulation from a fixed location
Analysis of home range from radio-location	Minimum polygon area; probabilistic methods; harmonic mean; Kenward (1987) 'Ranges' suite of programs

Figure 7.12

Comparison
between
radio-
tracked and
visually
marked
Pheasants

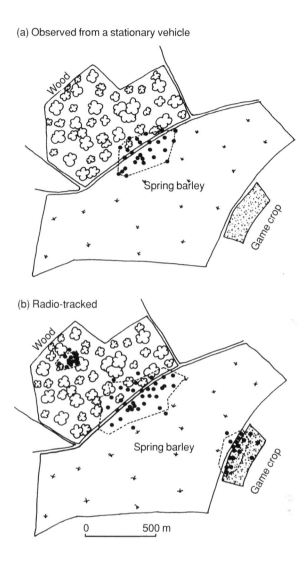

(a) Observed from a stationary vehicle

(b) Radio-tracked

Hypothetical example of the value of radio-tracking data when compared with
that obtained from the same bird which was back-tabbed. In (a) the data
obtained from a census from a stationary vehicle are restricted. If the
Pheasant (•) enters the wood, or moves a long distance, the census might fail
to pick it up. In (b) the same bird is radio-tracked, and this identifies an area
around the nest-site within woodland which was previously unrecorded, and a
feeding area in an old game crop a long distance from the main part of its
range. Consequently, if the stationary vehicle census data were used to say
something about habitat selection by the bird, a biased picture would be
presented.

Figure 7.13

Radiotagging
pheasant
chicks

Kenward et al. (1993) describe a method of radiotagging pheasant chicks in order to study survival during the critical first 14 days of life. Females are first caught before the breeding season and equipped with necklace-mounted radios. Nests of females are then found and broods located once hatched. Broods are radio-located to within 2–3 m accuracy on a dry night. On approach of the brood a torch is flashed on briefly to locate the brooding female and a wire mesh cylinder (20 mm mesh size) placed quickly over her. The cylinder has a more flexible 10 mm hexagonal mesh skirt which helps to prevent chicks from slipping out under the netting. The adult female is placed in a sack and the chicks transferred to a box containing chemical handwarmers. An observer remains near the roost site for 15 min to listen for distress calls from any lost chicks. Radios of 0.8–0.9 g with a 10-day life and range of 100–200 m are attached to the chicks within 5 days of hatching as (a) patagial tags, (b) backpacks with underwing loops, (c) glued with false-eyelash glue to the trimmed down on the back. Chicks marked with glue-mounted radios exhibit similar loss rates to untagged controls and produce better results than those marked with the other methods of attachment. Problems with the method include difficulty in distinguishing between signal loss due to trauma (e.g. destruction of tag with chick by a predator) and signal loss due to tag failure. Another is that it is difficult to determine the cause of death when a chick is found dead.

Summary and points to consider

Is it necessary to catch and mark the individuals in order to satisfy the objectives of the study? How is this related to bird counting?

Design of marking method, catching protocol and time constraints must be addressed.

If population size is to be estimated using catching and marking methods, are any of the assumptions of the analytical method violated? Can any violations be dealt with?

Consider a number of methods for achieving the same product, as in the Stock Dove example where long runs of data are available. Are these products similar, i.e. do the methods provide a consistent interpretation of, for example, changes in population size?

Wherever possible consider standardised and replicated procedures for collecting data from marked individuals. Reference, for example, to the Constant Effort Sites scheme is important here.

8

Counting Individual Species

Introduction

For some species the 'standard' methods of counting breeding populations; such as territory mapping (Chapter 3), point counts (Chapter 5) or
line transects (Chapter 4) are not particularly successful. Reasons for this
may be a low breeding density (e.g. raptors), a secretive nesting habit (e.g.
ducks, many waders), crepuscular or nocturnal ecology (e.g. owls,
nightjars), or semi-colonial and colonial nesting behaviour (e.g. herons,
Rook). However, specific methods have been developed for many of the
different types of birds where these problems occur. These aim to produce
either an index of the breeding or non-breeding population and thereby
facilitate comparisons of populations between years and sites, or count the
entire population as accurately as possible. In some cases the methods
developed to count these species are modifications of territory mapping or
transect methods, but other more novel methods have also been
developed for particular groups of birds, in particular habitats, or for a
single species. Some of the main groups of methods are described below,
with examples taken from the literature, and some highly specific
examples of methods are also outlined. More detail of single species
methods for species breeding in the UK is found in Gilbert *et al.* (1998),
but here the main aim of the examples is to illustrate methods that have
been grouped into similar categories.

Direct counts

Direct counts rely on being able to see the birds to be counted. If this is
possible, a suitable vantage point is selected and all visible birds are
counted using binoculars or telescopes. The method is very useful when
the birds are large and can be easily seen, e.g. swans or geese feeding on a
field.

With direct counting, a high level of accuracy at a given time may be
possible, but there are several ways in which the results can become biased.
The most important is a failure to ensure even survey effort and a similar
study area between years, resulting in data that are not comparable. Other
factors such as the weather during the counting, the people undertaking

the counts and whether the naked eye, binoculars or telescopes were used will all influence the accuracy and comparability of the counts. Some examples of the use of direct counting methods to count individual species in a variety of habitats are given below.

1. Great Crested Grebe

This is a wetland species which breeds on open waterbodies. The counting unit is the adult bird counted early in the breeding season. A recent study in Denmark (Woolhead 1987) indicates that populations can be estimated by counting the number of adult Great Crested Grebes on a waterbody several times, until the start of the first nest-building attempt, and then halving the average number of adults obtained from these counts. However, overestimation is possible on large lakes as non-breeding birds tend to congregate and might be regarded as breeding there. Counting early in the breeding season prevents problems with birds becoming obscured by vegetation around the edge of the waterbody.

2. Red Grouse

This is a species of upland dwarf-shrub heathlands. The counting unit is territorial encounters by males. Such encounters can be counted from late autumn to spring; during this period the males are on territory, frequently display and have numerous territorial disputes (Hudson and Rands 1988). Males can be counted from vantage points, or while the observer is walking or driving transects across suitable habitat.

3. Grey Partridge

This is a species of lowland farmland. There are two separate direct counting methods.

1. Spring count. The counting unit is the individual bird. These are counted in March, when the birds have paired and before the vegetation gets too high. Counts can be made in the 2 hours after dawn and the 2 hours before sunset, when the birds should be feeding in the open. Calm and dry conditions are preferred (Potts 1986). Birds are counted by an observer using binoculars from within a vehicle. Using this method, up to 200 ha of farmland can be covered in about 2.5 hours. Birds show up well in newly planted crops which can be counted from the field margins, but the vehicle has to criss-cross well-grown grasslands. On each visit maps showing the location of birds should be produced (Fig. 8.1). The maximum number of pairs plotted on the survey maps is used to produce a population estimate for the study area.
2. Post-breeding stubble count. The counting unit is the individual bird, or pairs with chicks. These are counted in the first 2 hours after sunrise during August. The post-breeding population can be estimated from the peak counts, and the breeding success can be quantified from the number of young birds observed and the proportion of pairs with no chicks.

Figure 8.1

Direct
counts of
Grey
Partridge

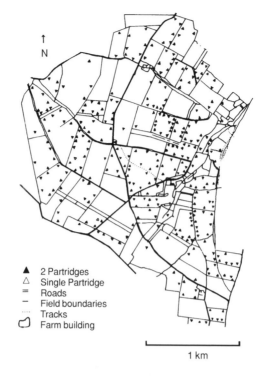

▲   2 Partridges
△   Single Partridge
=   Roads
–   Field boundaries
.... Tracks
⌂   Farm building

1 km

By plotting all the birds seen on an early morning survey of farmland between
mid March and late April, the number of Grey Partridge on the site can be
counted and some idea of the population and their habitat preferences
obtained (from Hudson and Rands 1988). For example, it is clear from this
figure that most of the Partridges are associated with field margins, with relatively few seen in the centres of fields, and also few alongside the road running
roughly north–south through the figure. Moreover, some fields appear to
support more Partridges than others and the birds seem to be approximately
equal distances apart in individual fields. Various hypotheses could be erected
to explain the population level and its distribution within the survey area; these
could then be tested on the ground. Similar surveys repeated after the
breeding season would assess productivity and could be used to see if there is
a shift in the positioning of the birds.

4. Lapwing

This is a species of short damp grasslands and agricultural fields. The
counting unit for Lapwing is the incubating bird. Counts are recommended when the birds are sitting on eggs (late March to late April in
Britain) because later juveniles, finished and failed breeders flock and
confuse the count (Reed and Fuller 1983; Barrett and Barrett 1984). Incubating birds are located by carefully scanning the study area. Several counts
should be made over the recommended period, between 09.00 and 12.00
hours British Standard Time (BST) when activity is most stable. Detailed
experimental work has concluded that the maximum of a series of counts
made during the period when most pairs are incubating gives a good

estimate of the number of birds breeding (Green 1985b). A recent survey of breeding Lapwings in Britain counted birds in a randomised selection of 10-km squares throughout the country using volunteers and the form shown in Fig. 8.2.

Figure 8.2

Survey sheet for the 1989 British Lapwing survey

AREA: Observers should count all nesting Lapwings in the tetrads (2 × 2-km squares) which have been selected at random by the National Organiser. All fields and other open habitats in each tetrad should be checked. Please return a card for every tetrad selected, even if apparently unsuitable or no Lapwings are found. **Nil returns are wanted.**

COUNTING: Lapwings are conspicuous when nesting and displaying. Territorial males, pairs or birds standing guard near nests are easily counted. On short vegetation incubating birds are easily seen and all fields should be carefully scanned for them. Counts of nesting pairs should be based on these points. Extensive open ground can be checked from roads, tracks and footpaths without disturbing nesting birds and this technique should be used whenever possible to ensure accurate counts. Even in apparently flat fields undulations can hide birds, so please ensure that all the field has been studied. Walking through a field (with permission), or a Crow flying over, will usually cause all the Lapwings to fly up, including any incubating birds. In this case the total seen may be halved for the number of pairs. **Please ask for access if it is necessary to enter private land away from public rights of way.**

TIMING: A complete count of each tetrad (which may need more than one visit to cover thoroughly) carried out in April is wanted. Nesting birds and nesting habitat being the aim, counts in southern counties may be best timed for the first half of the month, and in northern for the second half.

NOTES ON FILLING IN THE CARD:
1. Please record the total number of nesting Lapwings found in each habitat in each 1-km square in the appropriate box. If no Lapwings are found in a 1-km square put a zero in the totals box only. Any hatched broods seen should be recorded on a separate sheet.
2. Only record livestock if present during counts, noting C, S or H as appropriate, in the relevant box.
3. Farms may keep both cattle and sheep (and horses) and cattle and sheep may be grazed together. In these cases summarise the number of nesting Lapwings present with each type of stock in the relevant box as, e.g. 5C, 2S, 3 C+S etc.

4. Some autumn cereals are tramlined and some may have bare patches after the winter, which attract Lapwings. Where pairs nest in autumn cereals please note P (bare patches) or T (tramlines) if appropriate, e.g. 4T 5P 2 etc.
5. Record any pairs in oil-seed rape under Other habitats.
6. An enlarged copy of the 2½ inch OS map of each tetrad is supplied with each card. These show field boundaries. On the map please record the number of pairs found nesting in each occupied field and the field's boundaries. The latter is important as many of these maps are rather out of date.
7. **Autumn and spring cereals** differ sharply in April. Autumn cereals are taller and/or bushier and usually darker green. Many spring cereal crops will still be at the single leaf stage, having just emerged.
8. **Tramlines** are the permanent wheeltracks left across growing cereals.
9. **Other spring crops** are crops other than cereals planted in spring, e.g. potatoes, peas.
10. **Plough** is bare land still in furrow. Tilled land is bare land worked down fine but without an emergent crop. Please record bare land as one or the other, even if it is known to be planted.
11. **Permanent grass and ley** are often difficult to distinguish. Leys usually lack the varied plants, uneven structures and firm turf of permanent grass. Young leys left for forage may be very even and dense stands, resembling thickly-sown cereals rather than a grazed meadow.
12. **Don't know grass.** It is important to separate permanent grass and ley whenever possible, but when it is not please record the field under this heading.
13. **Rough grass** should be taken as poor grassland infested with rushes or weeds.
14. **Tetrads.** The lettering system for tetrads is illustrated right.
15. **Previous counts.** If you have any previous counts for this tetrad please enter year and number in the box provided or on a separate sheet.

E	J	P	U	Z
D	I	N	T	Y
C	H	M	S	X
B	G	L	R	W
A	F	K	Q	V

Figure 8.2

(continued)

SURVEY OF NESTING LAPWINGS IN ENGLAND AND WALES

British Trust for Ornithology

Please read instructions on reverse and fill in the map

For office use

County	10-km square no.	Main village or town	
Observer: Name	Tetrad letter	Village or farm	Dates of visits
Address		Was tetrad fully searched?	

NUMBER OF NESTING PAIRS OF LAPWINGS IN:

	Ley grass (note 11)	grazed C S H*	ungrazed
Autumn sown cereals (notes 4,7,8)	Perm-anent grass (note 11)	grazed C S H*	ungrazed
Spring sown cereals	Moor, rough grass (note 13)	grazed C S H*	ungrazed
Sugar beet	'Don't know' grass (note 12)	grazed C S H*	ungrazed
Other spring crops (note 9)	Reservoirs		
Bare land (note 10) Ploughed / Tilled	Sewage farms		
Stubble	Airfields		
Saltmarsh	Other (specify) (note 5)		
Dunes	TOTAL		
Gravel pits			
Waste ground			
TOTAL			

*C = cattle. S = sheep. H = horses (notes 2, 3)

Any previous counts? (note 15).

Please return by 30th June to Regional Organiser or to BTO. Beech Grove, Tring, Herts, HP23 5NR.

In this survey sheet, data on the number of birds recorded, their activity and habitat they were found in are requested. This enables both population and habitat studies to be completed. Results of this census are presented in Shrubb and Lack (1991).

Indirect counts

Indirect counts rely on counting signs of bird activity (droppings, burrows, etc.), rather than the birds themselves. This may be the only suitable method for particularly secretive (e.g. rain-forest pheasants) or non-visible birds (e.g. those nesting in burrows: Chapter 9). Theoretically, indirect signs of birds can be calibrated to produce indices of population level, for example when counting nesting burrows, or wildfowl (see below).

1. Waterfowl dropping counts

Determining the distribution and population of feeding waterfowl can be very difficult as they may roam in large flocks over an extensive area. A simple way to estimate populations using certain areas is to count faecal pellets (droppings) within plots (Owen 1971; Gibbons *et al.* 1996). However, when counting droppings, both the number of birds and the length of time they have spent in that particular area will influence the number of droppings recorded. Hence, the number of droppings counted per unit area gives an index of bird usage, not necessarily population level.

To use the dropping count method, first it is necessary to determine how long droppings will last before they disintegrate, or become indistinguishable from each other. Fresh droppings can be marked and then revisited over a period of time to study the speed of their disappearance. However, this is considerably influenced by rain and other weather conditions. Once the approximate rate of disappearance of the droppings is known, then Bamboo stakes are placed randomly (around 20 stakes) within the study (feeding) site, and the number of droppings is measured in a known area around these stakes. The simplest way to count the droppings is to tie a piece of string of known length to the pole and then tie a spoon to the end of the string. The spoon can be used to flick the droppings out of the circle as they are counted. The area is revisited before any new droppings will have disintegrated, enabling mean numbers of droppings per day to be calculated.

Counts of droppings can be converted to bird-days usage by estimating the dropping rate of the species concerned, which also gives an indication of the population level of the birds concerned. For observable species it is possible to assess the rate at which droppings are produced, by simply watching this process, but is much more difficult for species that cannot be observed.

In addition to larger waterfowl such as geese, this method can also be used for other species such as gamebirds. The method would also be suitable for other larger birds, particularly ground-dwelling species which have distinctive droppings.

Look-see counts

This method uses prior knowledge of the habitat preference of the species to be counted as the mechanism to make detailed searches for that species in likely locations. More than any of the previous methods, look-see counting requires the observer to 'know' well the bird concerned, particularly its habitat preferences. The first stage in the counting procedure is to locate study areas. Potentially suitable habitat for the species of interest is identified (1) from a map of the area, (2) from an aerial photograph or satellite image of the area, or (3) through contact with local experts. Once suitable areas are identified, a programme of site visits is arranged at the relevant time and using appropriate methodology to count any birds present. From such counts, population estimates are made (Fig. 8.3). Such counts will become biased if the area covered or the effort put into the counting varies over time. There is also a general tendency for more birds to be discovered over a period of years as the study area becomes better known, and thus for counts to increase regardless of changes in the actual population.

Data on the habitat preferences of many birds, which can be used as the basis of look-see counts, are available for Europe (Cramp and Simmons 1977, onwards; Bezzel 1985 and 1993), Africa (Brown *et al.* 1982), North America (Palmer 1962, onwards) and most other regions of the world.

Some examples of species which are counted using look-see methods are provided below.

1. Divers

For Red-throated, Black-throated, and Great Northern Diver, which breed on waterbodies in northern regions, the counting unit is the adult bird. The best counting period is late incubation or early fledging as all territories and nests are by then well established (mid-May to late June in Britain). The first step in the counting procedure is to locate potential breeding waterbodies within the study area on maps. Once located these sites are then visited twice during the breeding period, with the second visit not less than 2 weeks after the first. Two visits are needed as the birds may be feeding away from the site on one of the visits. On each visit the whole waterbody should be systematically scanned from suitable vantage points to detect adult divers on the water (Fig. 8.3), and all known nest-sites checked. Walking along the shoreline, or searches for nests, should be avoided as they are extremely time-consuming and may disturb the nesting birds. Detailed studies in northern Scotland (Bundy 1978; Campbell and Talbot 1987) used the following criteria to define a breeding territory:

1. chicks seen
2. nests with eggs seen
3. pair of adults on loch on two visits
4. pair of adults on one visit but single adult on the other visit
5. single adult on loch on both visits.

Figure 8.3

Look-see
surveys of
birds in
remote areas

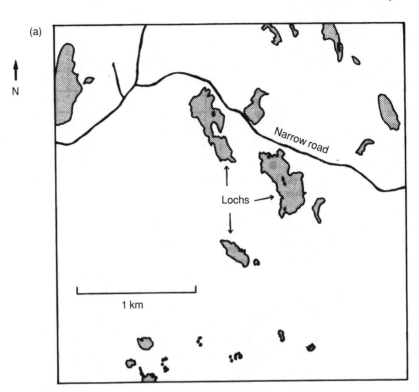

(a) When counting divers on lochans in Scotland all the lochans in the study area must first be located and a programme of site visits planned to each of them, giving equal coverage effort.

(b) At each site the whole surface of the water must be scanned slowly and carefully from side to side. Too little time spent at each site will lead to birds being missed if they are behind vegetation, during periods of poor visibility, etc. To obtain accurate results even the most remote and inaccessible lochans must be counted using the same amount of effort as the most accessible sites. Otherwise results might reflect the ease of reaching sites rather than the population of birds in the area.

2. Mute Swan

For this species, which breeds in wetlands and along rivers, the counting unit is the territorial bird with a nest. Counts should be made when the territories are well defined and the nests conspicuous (April or May in Britain). Males vigorously guard their breeding territories from all potential intruders and hence are easy to count. Proof of breeding is the location of an active nest or brood of cygnets. Potential bias in counts results from poor coverage of waterbodies, and because non-breeding males will sometimes set up and defend territories even though they are not paired.

The Mute Swan population of the whole of Britain has been estimated by counting territorial pairs in 10 km square counting units in April and May (Ogilvie 1986). Fig. 8.4 shows the census form used in the 1990 Mute Swan census of Britain.

Figure 8.4

BTO/WWF/ SOC Mute Swan census 1990

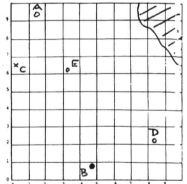

BTO/WWT/SOC MUTE SWAN CENSUS 1990: BREEDING PAIRS

County (England & Wales) or District & Region (Scotland)____AVON____

Please mark on the grid the positions of all pairs and nests found, using the following symbols to represent the state you recorded on your last visit:

Territorial pair .... ×
Pair with nest .... O
Pair with brood .... ●

Against each symbol write the letter used for the site code on the other side of this form.

Please shade any parts of the 10 km square that you were unable to cover.*

What is your best estimate of the number of pairs in the shaded part? What is your reason for this belief?*

O

No SUITABLE HABITAT

Please write your name, address, and phone no. here:

J. SMITH
I FIELD ROAD   654321
BRISTOL
AVON.   BSI IAA

*Note: estimates are much less useful than proper coverage.

As soon as possible after 31 May, please return to your local organizer.

Your local organizer is:

SIMON DELANY
WWT, SLIMBRIDGE, GLOS.
GL2 7BT   (0453-890333)

MUTE SWAN CENSUS 1990: BREEDING PAIRS: OBSERVATIONS

Site code	Location	Grid ref.	Habitat	Dates and observations (Dates as 00/0 please)
A	LITTLETON BRICKPITS, NR. WICK	716092	GRAV. PIT	03/4 T, 15/4 N, 01/5 N
B	ALVESTON RES.	747007	RESERVOIR	03/4 N, 01/5 N, 27/5 T
C	HILLSIDE LAKE THORNHAM	701065	LAKE	03/4 T
D	LITTLE AVON NR. THORNBURY	784021	STREAM	15/4 N, 07/5 D
E	R. SEVERN, FROME	733062	RIVER	01/5 N

3. Golden Eagle

This and other large raptor species breed in remote rocky areas, which are easily located on maps of a region. Once potential sites are found then a programme of site visits is undertaken to search for breeding birds, with care being taken to standardise the length of time spent at each potential site (Watson *et al.* 1989). The counting unit is the Apparently Occupied Nest-site, or Breeding Territory. Proof of occupancy by a pair would be:

1. seeing two birds together
2. finding moulted feathers or droppings, or preferably
3. finding a nest containing eggs or young, or seeing adults carrying food or hearing the begging calls of young birds.

4. Sparrowhawk

This species nests in woodlands and the nesting sites are generally quite evenly spaced in the landscape (Newton *et al.* 1986). Counting units are territorial birds with nest-clusters, and these should be counted in the early spring. Recommended methods for locating the nest-clusters are first to locate all potentially suitable woodland areas in the study region, then to visit these sites and look for

1. adult birds in or near a woodland
2. droppings, feathers from plucked prey or pellets on the woodland floor, and finally
3. the nests themselves (Newton 1986).

Because of the regular spacing of nesting areas (Newton *et al.* 1986), once one nest has been located a useful method to locate the next nest is to draw circles of between 0.5 and 2.0 km diameter on a map of the region (with the first nest as the centre) and then visit any woodland sites intersected to check for nests. The spacing of the nests varies according to the 'quality' of the habitat and hence when this distance has been identified in a given region, then locating new nests is made even easier as the appropriate circle diameter can be used to identify likely nesting sites much more rapidly.

5. Raven

This large and obvious species nests in remote areas very early in the spring. The counting unit for the Raven is the Occupied Territory. Territory occupancy can be determined early in the year (January/February in Britain) by assessing the number of birds at potential nesting sites. Ravens are conspicuous and their nests are built in easily located sites, are bulky and persist for several seasons. In the study of Marquiss *et al.* (1978), information on the number of Ravens in a study area was built up from a combination of local knowledge and active searching for nests. In general pairs are regularly distributed in an area of similar habitat, so by plotting the distribution of known pairs on a map, gaps become apparent which can be checked on the ground. Over a period of years the exact number of Ravens in an area can be determined.

Developments of look-see methodology into formally randomised population studies

In recent years attempts have been made in the UK to bring statistical rigour to surveys which go out and look for individual species, using a combination of information on the known range of the species concerned, its known habitat preferences, breeding period and detectability. Examples of the application of the method are Rebecca and Bainbridge (1998) for Merlin in the UK, Hancock *et al.* (1997) for Greenshank in the UK, and Underhill *et al.* (1998) for Common Scoter in the UK.

For all studies a formal randomisation procedure allowed population estimations to be generalised from incomplete survey coverage, and statistical confidence limits can be given to these population estimations.

The first stage in these studies is to define the breeding range of the species. In these papers recent breeding Atlases of the UK were used to do this. The breeding range was then stratified into strata where there was a high population (from the Atlas) and where the population was lower. The cut off levels for the high and low strata are selected by the study designers. Then survey squares (typically of 10 × 10 km or 5 × 5 km) were selected randomly. A larger percentage of squares from the areas with a presumed high population were chosen to assist making the final population estimate as precise as possible. Sampling intensity in the UK examples outlined above has been around 8% when averaged over the two strata.

On the ground the surveys were undertaken using detailed field knowledge of the species concerned, avoiding areas which were not favoured, concentrating on areas which were known to be favoured, visiting the areas at the best part of the season for finding the birds, for example. In this way the ground survey part of the work was quite like previous look-see surveys. The survey workers did not know the strata (high or low) the squares they had been allocated to count fell into, which further assisted the statistical design of the study.

Data from the breeding surveys were then turned into a population estimate for the species concerned following the methodology outlined in Fig. 8.5. This type of survey design and field and analytical methodology can be applied generally to different species and areas where a population estimate is required.

Vocal individuality counts

For some species it is possible to recognise individuals based on their calls and songs (e.g. Wickstrom 1982; Saunders and Wooler 1988; Gilbert *et al.* 1994). This technique is most useful for counting the individuals in small populations, and the effort is most worthwhile for rare and threatened species that cannot be censused easily in any other way (e.g. European Bittern in a British reedswamp).

This method is often the only way to be sure that bird calls are from different individuals, and also results in minimal disturbance, which is

Figure 8.5

Use of ran-
domised
survey
methodology
to improve
population
estimates

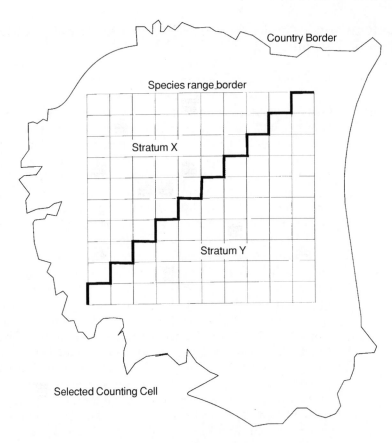

A common problem in field ornithology is the need to obtain a population
estimate for a single species of conservation concern (declining, increasing,
tied to a particular habitat, etc.). For such species surveys have been
conducted for many years over their breeding ranges using detailed local
knowledge of where the birds are and their habitat requirements. However,
such an approach lacks statistical rigour and hence the population estimates
cannot be given formal confidence limits. However, it is possible to make the
best use of local knowledge and field skills within a statistically solid
framework, so that field craft is used to obtain a statistically defensible population estimate of a species population. An example of the approach is given
below.

Here the larger circle represents the geographical borders of a country
within which a population assessment of a rare species is planned. The
gridded square represents the known distribution of the species to be
surveyed.

The grid represents possible survey squares. These are first stratified into two
regions, one (X) with high probable nesting populations (with this estimate
derived from an existing breeding bird Atlas), and another (Y) with low

Figure 8.5

(continued)

probable nesting populations. This stratification allows greater attention to be paid to the areas with a higher population which are the most important in the overall estimation of the population.

A random selection is made of the grid cells in strata X and Y and these grids form the survey areas for the teams of counters (generally volunteers). The populations of birds in these grid cells are counted using the accepted and standardised methods which make maximum use of field craft and field skill in the observers, who generally know the life history of the birds extremely well.

A population estimate is then derived from strata X of the range of the species using the following equation:

$N = nA/a$

Where:

N = the estimation of the species population status.
A = the area of the study region (in this case strata X of the range of the species).
a = the area of the sample cells where bird populations have been counted within strata X.
n = the number of birds counted in the sample cells within strata X.

The confidence limits on this population estimation are calculated as follows:

upper limit = $n+$ (mean + 1.96 × SE) × (A–a)
lower limit = $n-$ (mean + 1.96 × SE) × (A–a)

It is important to note that the confidence limits must be calculated for the whole study area **minus** the area where the study has been undertaken. If this is not done then incorrect results will be obtained.

important where rare species are being counted. However, the production of good quality recordings is hard work and requires a good deal of expensive equipment and further time and skill to analyse the tapes which are made, thus the method can only be used for rare species which justify the effort and cost.

Counts using a bird's association with another species

Some bird species associate strongly with other animals, e.g. in tropical forests there are ant-following birds and monkey-following birds. There are also birds in tropical forests which follow other bird species in multi-species flocks of quite stable species composition. Methods for counting populations for these species are not well defined and often have to be fitted to the particular circumstances, and the particular life histories of the birds concerned. Some examples of methods are given below:

1. Birds and monkeys

Various species of birds habitually follow monkeys, especially in the Neotropics (e.g. Boinski and Scott 1988). For some of these birds the best way to locate them and thus assess their populations in an area is first to locate the troops of monkeys that they associate with. In the study of Boinski and Scott (1988) the presence of birds in association with monkeys was recorded in a two-minute sample period every half hour, when the activity of the monkeys was also noted. The birds associated with the troops most during the wettest season, when arthropod abundance was lowest, and the birds probably benefitted by being able to find more food which had been disturbed by the monkeys. An indication of the abundance of a monkey-following bird can be assessed through the location and observation of the monkey troops.

2. Birds and ants

There are many species of birds in the tropics that follow colonies of ants and feed on invertebrates disturbed by the ants (e.g. Wiley 1980). It may be possible to get an idea of the population of some of these birds by observing ant columns over a period of time.

3. Birds and habitat features

Some birds are particularly strongly attracted to small habitat features, which if known and located can be used to make an estimate of their population. For example, in tropical forests flowering and fruiting trees attract birds which feed on these items from a considerable area. An idea of the presence/absence, and perhaps of abundance can then be assessed by spending time observing the birds coming to such trees. In many tropical forests some of the trees also grow much taller than others. When these trees die they become important staging posts for birds moving through the canopy and as territorial marker posts for other species. They have the advantage that birds sitting in them can be much more easily observed than if the tree is still alive. In many parts of the world birds are also attracted to small waterbodies to bathe or drink and individual birds are often faithful to particular sites. In areas with a pronounced dry season, such sites can be very important as locations where birds which cannot otherwise be observed can be found. There are certainly numerous other possible associations between certain bird species and particular smaller habitat features which could be used to make crude population estimates of species which are otherwise very difficult to count at all.

Nest searches

This labour intensive method can be used for any species, but has been most often used for birds of forests and shrublands in the USA (Ralph *et al.* 1993).

Nest finding can provide a direct measure of the population of an area, the success of breeding, and the choice of habitat for nest location. The method is most profitable in temperate latitudes where there is a well defined and short nesting period. Nest finding is slow, but the success can be improved through practice and training, especially by searching likely locations in the vicinity of an active female, and by starting searching during the nest-construction period. In detailed studies involving nest finding, it is recommended that one person monitors two plots of 40–50 ha within forest habitat, alternating the searches of these on alternate days (Ralph *et al.* 1993). The numbers of nests found can be used as an index of population level. Vegetation can also be measured at the positions of nests, and by also measuring the vegetation in random positions habitat preference and avoidance indices can be calculated (see Chapter 11).

Tree colony counts

There are a number of species which breed colonially in trees and during this period their populations can be counted quite accurately. This breeding strategy is most common in herons, egrets, storks, spoonbills, cormorants and some seabirds. Most species breeding like this are highly sensitive to disturbance of the breeding colony, especially early in the breeding season, and hence it is often not a good idea to attempt a census until egg-laying has commenced. Even then, extreme care should be taken to ensure minimal disturbance. For some species the tree nests are highly visible from a distance and hence counts of nests can be made from a suitable vantage point some way from the colony. An example of a more intensive method is presented below.

1. Grey Heron

This is a large wetland species which nests in colonies of 10s of individuals. The counting unit is the Apparently Occupied Nest. These should be counted in the late incubation or early nesting period (late April in

Britain), before the leaves of deciduous trees have emerged and obscured the view. Nest occupancy can be ascertained using the following criteria:

1. eggs in nest
2. eggshells beneath nest
3. young seen or heard
4. adults sitting
5. fresh nesting material found
6. droppings on or below the nest.

A mirror on the end of a long telescopic pole may also assist in establishing nest occupancy, although this equipment is heavy and may require two people (one to hold and the other to look at the mirror through a pair of binoculars). It may be difficult to distinguish between individual nests and in such cases the number of Apparently Occupied Nests must be estimated. Detailed counts in a sample of heronries in Scotland have shown that these methods record about 70% of the pairs using a site (Marquiss 1989). It may therefore be possible to produce correction factors for heronry counts.

In large study areas, all heronries, or a random sample of them, may be counted to produce an accurate picture of changes in the breeding population. Colonies naturally increase and then wane, with new smaller colonies being set up. Hence, in long-term studies of a regional heron population the new colonies must be located, otherwise counts will be biased towards showing a population decline at the traditional sites.

A method which causes no disturbance to the colony is to count the number of nests after the breeding season, but it may be difficult to assess if a nest has been used or not.

Soaring counts

Some species, especially large raptors, soar when they are searching for food and hence at these times they are quite visible and easy to count. Such counts can be made during both the breeding and the non-breeding seasons. Examples of counts of soaring birds are given below.

1. Car counts of raptors

In North America non-breeding raptor populations are assessed over large areas by counting soaring raptors from cars. The method involves driving slowly (17–40 km/h) on calm and clear days and counting all birds that one or two observers detect, usually within a specified distance (0.4–1.6 km) on each side of the road (Fuller and Mosher 1981; Ellis *et al.* 1990; Chapter 4). The method is good for the obvious soaring species, but cannot detect the more secretive forest-dwelling birds. Variability in the detectability of raptors, and consequently errors in density estimates derived from these counts, are discussed in a number of papers (Millsap and LeFranc 1988; Sauer and Droege 1990; Hanowski and Niemi 1995).

2. Common Buzzard

The counting unit is the soaring bird. These should be counted when they are on their breeding territories and when the ground has warmed sufficiently to allow soaring. The British breeding population has been assessed by counting birds in the early spring within randomly located 10-km squares throughout Britain (Taylor *et al.* 1988). Population estimates were produced by assuming one soaring bird was equivalent to one pair.

Playback counts

An assessment of the population of some bird species can be made by playing their calls from a tape recorder and then listening for an answering call from a live bird. These methods are most commonly used for nocturnal species, but can also be useful for species in dense wetlands, or dense forests, or where the species is particularly shy and secretive. Normally the records obtained are mapped, but they can also be recorded along transects or at randomly located playback points. Population estimates can be made in the same way as outlined in Chapters 3–5.

1. North American road counts

In North America, populations of owls along roads are counted using tape-recorders (Fuller and Mosher 1981). The calls of several species (starting with the smallest) are played at fixed distances (0.4–1.6 km) along transects defined by roads and return calls are listened for. The taped calls are repeated several times to allow the real birds to reply. A similar method is also used for raptors in America, and contact rates are significantly improved when playback is used (Mosher *et al.* 1990). A recent study of a single species of owl in America (Burrowing Owl), found that 53% more birds were recorded when playback was used on a transect than when it was not (Haug and Diduik 1993).

2. Stone Curlew

This species is typical of very short and open grasslands and agricultural fields of light and often sandy soils. The counting unit is the incubating bird. These are counted at dusk and during the night. Birds are located by playing tapes of their call from a slowly moving vehicle. If the taped call is within 500 m of an incubating bird it will answer and can thus be counted.

3. Storm Petrel

This species nests in burrows on offshore islands. The use of tape playback to assess the populations on Shetland has been recently assessed (Ratcliffe *et al.* 1998). The peak nesting period was discovered and the number of birds responding to playback was assessed against the number of nests known to be occupied (from detailed sample surveys). In general it was found that around 30% of the males in nests known to be occupied responded to tapes played into the nest-hole. This percentage varied between colonies, at different times of day and over the breeding season and hence use of the method needs to be accompanied by other detailed assessments of the colony.

A variation of the playback count is where the observer makes an imitation of the territorial song or call of the bird being counted. Examples of this technique are common among hunters stalking bird prey, and ornithologists trying to locate rare bird species. In the study by Swenson (1991) hunters' whistles were used to imitate the calls of Hazel Grouse, which then responded and could thus be censused.

Roosting counts

There are some large bird species (e.g. raptors, parrots) which are widely dispersed at low density, but which roost communally in roosts typically of a few tens to a few hundred individuals. Such species are best counted at their roosting sites.

1. Hen Harrier

The population of wintering Hen Harriers in Britain has recently been assessed using roost counts (Clarke and Watson 1990). Firstly the positions of roosts or suspected roosts was established by sending questionnaires to local conservationists and amateur ornithologists throughout the country. Then, a coordinated programme of watches of all roosts was organised and the number of birds present in mid-winter (January) was counted. This gave an estimate of the total population for the whole country, although in areas with fewer ornithologists there may have been more roosts which were missed, leading to some biases in the results.

2. Parrots

In South America one method for assessing populations of larger parrot species is to first locate the roosting sites, and then to count the birds as they arrive in the evening to roost. Such roosting areas are thinly spread as the parrots from a single site can range over extensive areas and hence most time and effort needs to be put into locating the roost sites through contacts with local authorities, ornithologists and naturalists, etc. In Britain the breeding population of the naturalised Ring-necked Parakeet has recently been assessed using a coordinated survey of all the known roosts (Pithon and Dytham 1999). A further assessment of different methods for counting parrots is provided by Casagrande and Beissinger (1997).

Dog counts

Some types of dogs (especially Pointers) are very good at finding birds nesting in dense (but short) vegetation. Using these dogs can be an efficient way to locate birds in a defined area and hence to assess their populations. The method is typically used for gamebirds, but also works well for ducks and other larger ground-nesting species. Some examples are given below.

1. Red Grouse

This is a species of upland dwarf-shrub heathlands. The counting unit is the individual Red Grouse, both territorial and non-territorial. The method is for the observer and dog to methodically search an area for a defined time period and locate the birds. Maximum detectability by dogs is in the autumn (October to December) and spring (March and April) (Hudson 1986). The total number of birds in the study area can be assessed using this method, but it is difficult to define the number of birds holding breeding territories.

2. Capercaillie

This is a species of upland coniferous woodlands. In this case the counting unit is the female with chicks. Counts should be undertaking during July when females are moving around with their broods (Moss and Oswald 1985). Females are counted by two persons with pointing dogs quartering the study area or walking transects 20 m apart; the dogs should locate all the birds and the presence of chicks can then be ascertained and mapped. If the number of females with broods has been assessed from a number of sample plots in a larger site the density of breeding females (and their breeding success) can be extrapolated for the whole site.

Shooting counts

Useful information on population fluctuations and trends can be derived from shooting bag records (Fig. 8.6: Tapper 1989). In Britain, there are long series of data available, from the large shooting estates, on birds such as Red Grouse. In some cases these data go back into the 1800s but as they are influenced by several factors apart from population level, their main use is in long-term studies of population dynamics (e.g. Barnes 1987). In many countries there are also detailed records of numbers of all types of waterfowl, waders and gamebirds which have been shot. All such data can provide an indication of population changes over time, especially over long periods.

Lekking counts

About 150 bird species congregate in communal display arenas, called leks (Johnsgaard 1994). At such places the numbers of males and females can be counted directly, with the key to accurate counts being to ensure that all the lekking sites in a study area have been identified.

1. Capercaillie

The counting unit is the male and/or female at the lekking site. The birds

Figure 8.6

Shooting bag
records

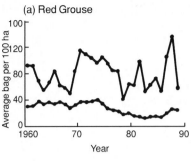

(a) Red Grouse

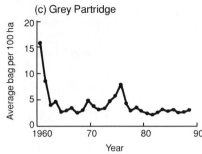

(c) Grey Partridge

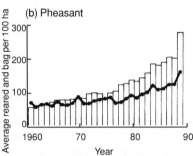

(b) Pheasant

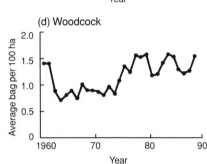

(d) Woodcock

Historical records of shooting bags of Red Grouse, Pheasant, Grey Partridge and Woodcock are available from a wide sample of sites throughout Britain (from Tapper 1989). These provide valuable information on population trends over extended periods. However, other factors need to be considered for each species before the true meaning of the graphs can be assessed.

(a) Average bag records of Red Grouse per 100 ha of moorland from several sites in England and Scotland. Numbers shot in England show cyclicity, with 1989 being a poor year. However, there are several other facts needed before the graph can be fully evaluated; these are the number of persons shooting, changes in the shooting equipment, periods of adverse weather, changes in the law on shooting or gamekeeping, etc. Such factors will influence the numbers of birds shot and hence the shooting bag records obtained.

(b) Average bag records of Pheasant per 100 ha. Taken alone this graph might tend to suggest an increasing natural population. However, the bar charts on the graph present the number of young birds artificially reared and released into the wild. Considering these data, it now appears that the number of birds shot has, in fact, not increased in line with the number released. An increasing number of birds being released also implies increasing interest in Pheasant shooting and thus increased shooting pressure. Much additional background data on trends in Pheasant rearing, Pheasant shooting and the status of the wild population need to be available before this graph can be fully interpreted.

(c) Average bag records of Grey Partridge per 100 ha. The graph shows a dramatic decline in numbers shot in the early 1960s, followed by an extended period when the number of birds shot has remained low and relatively stable. The population decline in the 1960s is believed to be due to the increased

Figure 8.6

(continued)

usage of pesticides from that time, reducing the food supply for Grey Partridge chicks (Potts 1986). However, to interpret the graph fully, the level of shooting pressure over time and the number of birds released over time would need to be known.

(d) Average bag records of Woodcock per 100 ha. The number of Woodcock being shot per annum has been increasing since the mid 1970s. However, this does not necessarily reflect an increasing Woodcock population. There may be, but as most Woodcock are shot on Pheasant shoots, the increased shooting of Pheasant may have also increased the bags of Woodcock.

should be counted from a suitable vantage point at dawn in the early spring (mid- to late April in Britain; Moss and Oswald 1985). Lekking sites can be identified by systematically searching suitable areas, looking for tracks in snow, or field-checking information provided from local sources (Rolstad and Wegge 1987). If all lekking sites have been located, an accurate assessment of the breeding population can be made. If it is unknown whether all leks have been recorded, then the mean of the maximum number of birds (cocks and hens) seen at lekking sites may be used to produce an index of the breeding population.

2. Black Grouse
The counting unit is the male bird. Peak numbers of males are found at the lekking site just before egg laying and just after dawn (Cayford and Walker 1991). As there is normally little exchange between sites then a single count of the males will produce a good index of the population. However, it is good practice to count sample leks throughout the day and the season to determine how attendance varies. Locating leks is the key to a good count and the calling birds assist this process. New small leks may be established as traditional leks decline and hence locating the new sites is very important if counts are not always to show a decline, regardless of the actual situation.

Driving/flushing counts

Many large and primarily ground-dwelling birds, especially gamebirds, can be counted if they are flushed by people, using the same methods as employed by hunters.

1. Capercaillie
The counting unit is the individual bird. Several people (beaters) arranged in lines 20 m apart, walk slowly through suitable habitat to flush the birds. The method is best used from late October to early November and gives an estimate of total numbers of birds, rather than breeding pairs (Lindén and Rajala 1981; Moss and Oswald 1985).

Very similar methods can be used for other species of gamebirds, for ground-roosting nocturnal birds such as nightjars, and for other species which will fly up from the ground when disturbed sufficiently.

2. Nesting ducks

In America, breeding ducks have been counted by flushing the birds from their nests. The method involves driving two jeeps with a 50 m cable chain between them through the grassland along a transect of known length. The number of females flushed per unit area is counted and used to assess densities (Duebbert and Lokemoen 1976). A second method utilises light aircraft to fly transects and count ducks flushed from their nests (Bellrose 1976).

Nocturnal and crepuscular species counts

There are particular problems with assessing the populations of birds which typically have a nocturnal or crepuscular ecology. However, for many such species, methods have been developed to count the bird's population during the breeding season, and occasionally at other times of the year as well. Some examples are given below.

1. Woodcock

This is a species of lowland areas, which spends the day in woodland and feeds on fields during the night. There are two counting methods.

1. Counts of displaying males. The counting unit is the displaying 'roding' male. The counting period is throughout the breeding season (April to end June in Britain), with the maximum activity in Britain occurring towards the end of May (Hirons 1980). Populations can be only roughly estimated because some males rode more than others and the birds do not occupy discrete areas.
2. Nocturnal feeding counts. The counting unit is the individual bird. These are counted as they fly to or from nocturnal feeding areas at dusk or dawn. Counts can be undertaken throughout the year and allow indices of population level to be produced.

2. Tawny Owl

This is a widely distributed species of woodlands. The counting unit is the territorial male. Counts are made in late autumn (October to December in Britain) when territorial activity is most intense (Southern and Lowe 1968; Mead 1987). All records of 'hooting' and other calls should be plotted on maps during biweekly visits to the study area. Territories are identified from boundary disputes between males and clusters of records as with normal territory mapping methods (Mead 1987; Fig. 8.7).

Figure 8.7

Specialised
mapping of
Tawny Owl
territories

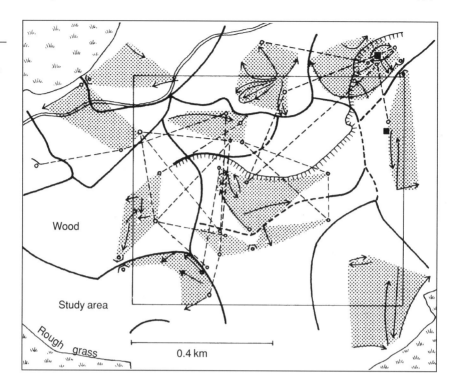

The figure presents Tawny Owl territories in Wytham Great Wood in 1954 and 1955, as determined by observation (from Southern and Lowe 1968). The large square is the detailed study area. Small circles connected by dashed lines show positions of birds giving territorial challenges ('hooting') simultaneously and thus separated. Brackets show territorial boundary disputes. The shading indicates the minimum area of a territory from the observation of territorial challenges. Black lines with arrows show observed movements and direction. Heavy black lines show the best fit of territories determined from observations and prey items marked with rings and recovered from pellets. The interrupted parts of these heavy lines show where changes of boundaries occurred before the year 1955/56. The black squares represent nest-boxes used for breeding and the inner triangle is the trapping grid.

This method is similar to the CBC but only territorial disputes can be used to separate territories. Interpretation of the results can produce density estimates for this species and can allow the position of territories over time to be mapped, the mortality and longevity to be investigated and the habitat relationships of the birds to be understood.

Reproduced with permission of the Institute of Terrestrial Ecology.

3. Barn Owl

This a widely distributed species of open habitats, typically farmland. The counting unit is a bird on its territory. The breeding population of Britain has been assessed by widely distributing questionnaires on the bird, allied with publicity campaigns, local knowledge and interviews with farmers (Bunn *et al.* 1982; Shawyer 1987). Only the proof of a bird on its territory has been required to define a breeding pair. In more detailed population studies Barn Owls have been counted by rigorous searches for nests. The procedure is to mark down on detailed maps of the survey area the location of all buildings, groups of trees, etc. that may provide suitable nest-sites. All potentially suitable areas are then rigorously cold-searched for nesting birds. In simple areas with few potential nesting sites this method is practical, in other areas it may prove too time consuming.

4. European Nightjar

This is a species of lowland heathlands and open coniferous woodlands. The counting unit for this species is the calling (churring) male. These should be counted at dusk on calm days (Beaufort force 0–4) throughout the breeding period (May to July in Britain). The number of churring males is assessed as the maximum number of birds that call from separate locations at dusk and for the next half-hour. Separate locations are generally defined as sites over 500 m apart with calling less than 30 seconds apart (Cadbury 1981; Gribble 1983). Large numbers of people are required to record individual calling locations on a large site and thus produce accurate population estimates.

Dense grass-dominated vegetation species counts

In many vegetation-rich wetlands and tall grasslands it is impossible to use the more normal counting methods, such as transects or point counts, owing to problems with access and visibility. Some population assessments can, however, be undertaken using transect methods from aeroplanes (e.g. Shoebills in eastern African swamps, ducks on North American Prairies), although this depends on the birds being large, flushable, and easily identifiable from a plane. Examples of more specialised methods which have been used for some of the other species which breed in dense wetland and dense grassland habitats are presented below.

1. Water Rail

This is an extremely secretive species of densely vegetated wetlands. A recent detailed study has used baited Potter traps in a small wetland to capture the Water Rails found on the site. This approach captured eight Water Rails in a 0.5 ha wetland (Jenkins *et al.* 1995).

2. Little Grebe

The Little Grebe lives in swamps and other types of wetlands, where it is quite secretive during the breeding season. In addition, pairs often make several breeding attempts, moving between lakes when doing so, which makes breeding populations difficult to assess. Vinicombe (1982) recommends making several counts of a study site during the first breeding attempt of the year (April to May in Britain). On each visit all sight and calling records are marked on a site-map. Often the only clues to the presence of breeding pairs are the 'trilling' calls.

3. European Bittern

This Bittern inhabits extensive reedbeds, where it is secretive and not possible to see with any regularity. One counting method involves locating and mapping vocalisations (booms) over the breeding period. Booming is of greatest intensity in the early morning and dusk, hence survey visits are recommended at these times. At least three survey visits should be undertaken to assess population levels, with the number of booming males being assessed as the number of clusters of booming records. This method has been shown to be somewhat inaccurate, and a more accurate count of Bittern populations can be made by recording and then electronically analysing the boom-patterns of all birds in a study area. As the sonographs are specific to a single male an accurate count can result (Gilbert *et al.* 1994).

4. Ducks

Breeding populations of many *Anas* and *Aythya* duck species are difficult to count as they nest in dense vegetation and often move their broods to other areas as soon as they hatch.

Three counting methods are commonly used.

1. Counts of nesting females. The counting unit is the female with nest. The most accurate method of locating nests involves rigorous searches in suitable habitat (e.g. Hill 1984a,b). However, foot-searches for nests are extremely labour intensive and may result in nest desertion; consequently they are rarely undertaken. Different species of duck should be counted at different times during the spring. For example, in a climatically average year in southern Britain, Mallard start nesting in March/April, Shoveler in late April/early May, Teal in April, Pochard in April and Tufted Duck in May. However, these timings can be shifted forwards or backwards according to the weather in a particular year, hence the best counting periods for most ducks will vary by year. Pöysä (1984) recommends a series of counts spread throughout the potential breeding period to identify the optimal census period for each duck species in a particular year. This time is recognised as when the maximum number of males have flocked together, after the wintering flocks have dispersed and before any post-breeding flocks return.
2. Counts of duck broods. The counting unit is the female duck with young brood. Counts can be made by direct observation of a site over a designated period, or by flushing broods onto the open water by walking the banks with dogs (Rumble and Flake 1982). Flush counts are

generally more successful and quicker than observations, except on larger or more vegetated waterbodies.

3. Counts of individuals and groups. This method has been developed and tested in Finland (Pöysä 1996 and references therein) and involves the use of transects to walk within 100 m of any point in a study area. All individuals and groups are recorded on a map or notebook. Large groups should be included, but care must be taken not to include non-breeding birds or late winter flocks. Counts can be interpreted in the following way for Wigeon, Gadwall, Teal, Pintail, Garganey, Pochard and Shoveler. Breeding pairs are recorded for all pairs, single males, males in groups of 2–4 (= 2–4 pairs), small male groups chasing a single female (2–4 male and 1 female = 2–4 pairs) and lone females if their number is larger than that of males.

5. Corncrake

This is a species of wet grasslands and hayfields which have not been agriculturally improved. Methods of counting Corncrake have been developed in northern Scotland and Ireland (Cadbury 1980; Stowe and Hudson 1988, and refined in Green 1995). The counting unit is the breeding territory. Calling birds (presumed to be mainly males) are counted on several visits to the study area between 23.00 and 03.00 hours BST in late May, June or even early July. The location of calling birds is plotted on a map of the study area on each visit. Playbacks of the craking call may be used to stimulate the birds into calling. The number of breeding territories is assessed using methods outlined in Chapter 3. A territory is regarded as being occupied if the bird is heard calling for more than 5 days (Stowe and Hudson 1988).

6. Snipe

This is a species of wet grasslands grading into swamps. The counting unit for this species is the displaying (drumming) male. These should be counted on at least three occasions during the display period (April and May in Britain), and within 3 hours of dawn or dusk (Green 1985a). Rope-dragging experiments, which discover all Snipe nests in a study field, have indicated that the true nesting population can be calculated by doubling the mean of April/May counts of drumming birds (Fig. 8.8).

7. Redshank

This is a species of wet grasslands and saltmarshes. The counting unit for Redshank is the flying bird showing alarm. These are best counted when the birds have young (late May to early June in Britain), observer traversing a field or walking set transects, and both should be counted individually. The maximum number counted approximates to the number of nests present (Fig. 8.8). Several visits spread over a couple of weeks are recommended to increase the accuracy of the results. However, the population can be estimated by halving the number of flying birds recorded on a single survey visit.

Figure 8.8

Specialised
counting
methods for
breeding
waders

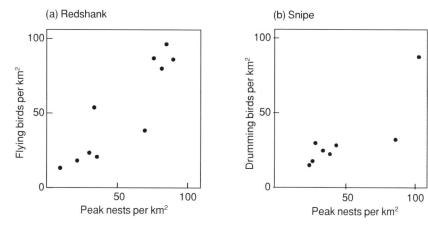

These graphs show the relationship between counts of birds per km² and peak numbers of nests per km² for (a) Redshank and (b) Snipe (Green 1985b). Each point represents the data from an individual study area. Peak nests per km² were calculated by locating nests by intensive rope-dragging experiments. Peak numbers of birds per km² were calculated somewhat differently for the two species. For Redshank, the mean density of birds seen per visit in April and May was used, and for Snipe, the mean density of drumming (aerial display) birds in April and May. At least three counts and usually five or six were made for each species. Fitted regression lines to these graphs (not shown) indicate that one Redshank counted in April and May represents approximately one nest at the peak period. For Snipe, one drumming bird is equivalent to about two nests at the peak time, although there are rather few data to substantiate this latter relationship.

8. Dunlin

This is a species of wet upland mires. It is an extremely difficult species to count accurately as it breeds semi-colonially, has small territories, does not move far to mob intruders, and is very inconspicuous (Reed and Fuller 1983; Reed *et al.* 1984; Brown and Shepherd 1993). The counting unit is the individual bird. These should be counted along transects during the day (between 09.00 and 17.00 hours BST in Britain). Greater detectability in Scotland has been found in June, as once the birds are incubating they are more difficult to flush and are even more subject to underestimation.

9. Curlew and Whimbrel

These are species of wet grasslands and mires. For these species the counting unit is the displaying bird. These are counted on three visits to the study area in the late incubation/early fledging period (late May/early June in Britain). Birds are counted while the observer walks transects 200–400 m apart, or from the edge of a field. All birds showing signs of being on territory are marked onto 1:2500 scale maps in the field. For Whimbrel, population estimates are produced from the identification of breeding territories based on criteria such as

1. alarm calls most intense
2. both birds of a pair calling actively overhead or undertaking distraction displays
3. one of the pair adopting a characteristic secretive 'creeping' run
4. both birds of a pair alighting near observer and calling in a highly agitated state (Richardson 1990).

Summary and points to consider

Individual species can be counted in a variety of ways, even if the more traditional methods are not well suited to assessing their populations.

Direct counts of some species are a relatively easy way to count the breeding or non-breeding population.

Indirect counts of droppings or other signs of birds provide a method for counting some species without ever observing them.

Look-see population surveys involve studies of areas of habitat thought to be suitable for the study bird, usually during the breeding season. It is important to be familiar with the study area and the ecology and behaviour of the bird concerned. Similar levels of effort should be expended in counting each potential site to avoid counts reflecting effort rather more than number of birds.

A formalisation of the look-see counting approach involves the use of stratified and randomly selected counting units (grids), within which the counts are undertaken. This randomisation of the survey design allows an overall population estimate to be calculated from incomplete data and also for confidence limits to be calculated for these estimates.

Methods which rely on the bird being surveyed responding to its call (tape playback) are efficient at locating and counting secretive, nocturnal or low density species. A modification is to record the calls of rare birds and analyse these using computers to assess the number of different individuals – which is particularly valuable for species which inhabit dense wetlands where they cannot be observed.

Various methods of locating nests, from direct searching, to flushing the birds from their nesting sites, to using other animals (dogs) to locate the nests through smell are all effective ways to locate ground-nesting species.

Drives with beaters are a useful way to count larger ground-birds which can be scared into flight. If the area beaten is known, then the density of the bird can be calculated.

Shooting bag records provide an index of the population level of hunted species. Bag records are, however, also influenced by factors other than population, such as shooting pressure and changes in methods (e.g. better guns).

There are still species, e.g. very abundant species found in extensive wetlands, such as Bearded Tit, or various swamp weavers, for which there is no real way known to assess their populations in anything more than a very crude way. These species pose a challenge to future ornithologists to devise methods which can be used in the field to provide a repeatable measure of the populations.

9

Counting Colonial Nesting, Flocking and Migrating Birds

Introduction

Counting colonially breeding birds (particularly seabirds) and non-breeding flocks of birds presents special problems, which must be addressed if accurate counts are to result. In a seabird colony these problems include the difficulties of assessing the proportion of breeding and non-breeding birds, locating and counting breeding colonies on remote and rugged coastal sites, evaluating the proportions of birds that have left the nest to obtain food, and defining the effects of harsh weather on numbers of birds at the colony. In flocks, the problems include the limitations of binoculars and telescopes, and the variability in the ability of observers to identify species of birds in flocks or to estimate numbers of birds within a flock, especially when several species of different sizes are intermingled. Despite these problems, however, procedures have been developed to estimate populations of birds when they are in breeding colonies, or in flocks. This chapter is divided into two parts, which discuss these two categories of bird counting.

1. Counting breeding colonies

Methods used to count populations of birds in breeding colonies are here divided according to the breeding habitat of the species: those that breed on cliffs, those that breed on the ground, and those that breed in trees. The majority of species breeding in such colonies are seabirds and examples of the methods used for specific species within the major categories are presented wherever possible. Methods of monitoring populations of birds in the breeding colonies are also outlined as breeding bird monitoring programmes are important in many parts of the world (e.g. Walsh *et al.* 1995).

Preparatory stages

Before counting the birds at a breeding colony several preparatory stages are recommended.

Figure 9.1

Division of
study area
into counting
sections
(from Bibby
et al. 1992)

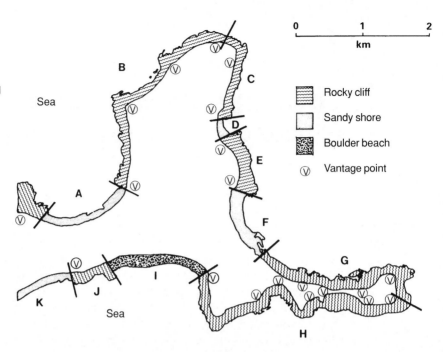

When colonies of seabirds are distributed along extensive seacliffs or shorelines, it is necessary to divide the cliff or shore into easily countable sections.
These are best defined by the features of the area (vertical cliff, boulders,
sandy beach, etc.), the availability of suitable vantage points from which the
birds can be counted, and because each section can be counted easily in one
go. It is important that all sections and vantage points are marked on the base-
map of the study area, and the results of the counts are presented according to
the various sections. It is also worth considering counting the sections in the
same sequence, or randomly, to minimise bias caused by colony attendance
altering over the counting period.

1. Description of study area

The region to be surveyed should be visited and the position of all colonies
and other breeding areas marked on a base-map at approximately 1:10 000
scale (Fig. 9.1). An alternative is to use aerial photographs as the base-map
to mark the locations of colonies. If colonies are spread along an extensive
length of cliff, or one colony cannot be viewed from one site, or if breeding
density is high, then the study area should be divided into counting
sections dependent on the availability of suitable vantage points.

2. Description of breeding colony

A seabird colony is a breeding assemblage of birds in a single location
where the individuals are close enough to interact socially (Gochfield
1980). If in doubt it is usually best to sub-divide a colony, so long as this can
be done unambiguously. For each seabird breeding colony the following
should be recorded (after Seabird Group/NCC 1988; Lloyd *et al.* 1991).

1. Colony name. Names should be the same as on a national map of the study area.
2. Location. These should be descriptive, e.g. north side of Firth of Forth, near Crail. Grid references should also be given (for the start and finish of the section of the cliff or the centre of the colony).
3. Status, e.g. National Park, Nature Reserve, private landowner (specific owner if possible).
4. Description. Details of site, including geology and vegetation. If possible, sketch these details in the field and take Polaroid photographs as a permanent record (writing date and details of colony on back of photograph). The location of counting positions and direction of view should also be marked (Fig. 9.1). It is important that the boundaries of the colony or sample plots are shown in relation to the main features of the region, streams, gullies, etc. so that they can be located exactly in the future.
5. Access. How to get to the site, landowner's name and address, etc.
6. History. Counting history if known.
7. Counting problems. Indicate any particular counting problems, e.g. birds nesting in caves, counted whilst looking up, broad ledges hiding birds, restricted view of colony, disturbance of colony by observer.
8. Other notes. Any relevant information on the colony, e.g. site of annual population monitoring.
9. Bibliography. Any details of books, scientific papers, reports, etc. that mention the colony.

3. Selection of counting method

The aims of the count and the species present will largely determine the methods used. Simpler and cruder methods (e.g. photographs) may be acceptable in the first survey of remote seabird colonies, but much more precise and time consuming methods (e.g. randomised quadrats) are required in detailed monitoring programmes. Some of the main categories of counting method are outlined below, with examples of the use of the method for various species provided where possible.

4. Assessment of risk to counters

As many breeding colonies are located on cliffs, the safety of the people doing the counting should be considered carefully. A number of experienced ornithologists have fallen to their deaths while counting cliff-breeding colonies; hence the dangers should never be underestimated and all precautions should be taken to minimise them. For example, at regularly used observation points, fixed ropes attached to the belt and braces of observers by carabiners may be necessary to prevent accidents. Mountaineers can provide valuable advice on sensible safety precautions on cliffs.

Counting methods for cliff-nesting seabirds

Detailed descriptions of counting methods for various seabirds are provided below. The bulk of the information has been taken from a relatively small number of publications (Nettleship 1976; Birkhead and Nettleship 1980; Evans 1980, 1986; Seabird Group/NCC 1988; Lloyd *et al.* 1991; Komdeur *et al.* 1992; Walsh *et al.* 1995; Gilbert *et al.* 1998).

Counts of birds on cliffs can be made within sample plots, which can be positioned in a number of ways. To obtain the most accurate counts at cliff-nesting colonies the location of the observer is important. Ideally, observers should be at the same level, or slightly above, the birds and should be looking directly at the colony (Fig. 9.2). If this preferred position cannot be obtained, the observer will be forced to count the birds from available locations. As outlined above, observer safety should be considered an overriding priority in the selection of the counting position.

Figure 9.2

Positioning of observer

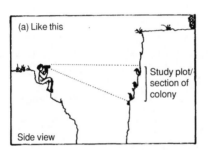
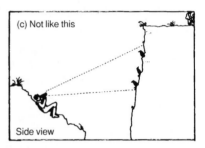
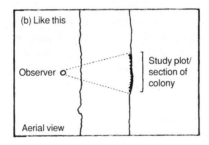
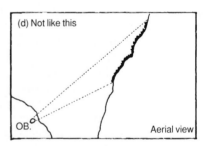

Correct and incorrect positionings of observer for viewing and counting study plots or whole colonies of breeding seabirds are shown (from Birkhead and Nettleship 1980). (a) Side view – observer should be slightly above breeding birds; (b) aerial view – observer should be directly opposite the study plot. In examples (c) and (d) poor positions for counting birds are presented. However, observer safety should always be considered the number one priority and if the ideal positions for counting birds cannot safely be obtained, then a compromise should be reached.

Figure 9.3

Techniques for positioning sample plots along a seabird cliff (from Walsh *et al.* 1995)

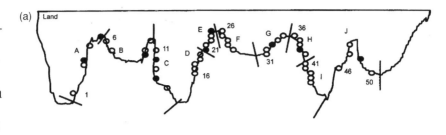

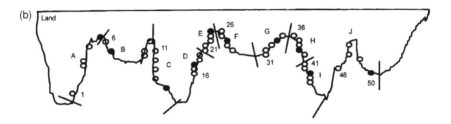

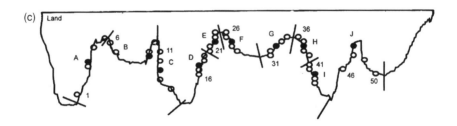

In this hypothetical example a preliminary survey has identified 50 suitable, potential plots, and 10 larger sections (major divisions), each holding similar numbers of birds. Plots to be counted on 5 to 10 dates annually, and, in the time available on each date, 10 study-plots can be counted.

a) Unrestricted random positioning. Ten plots positioned randomly along the coastline as a whole (regardless of the major divisions). Plot numbers are selected randomly using tables of random numbers, or a calculator.

b) Stratified random positioning. One plot positioned randomly within each of the ten major divisions.

c) Systematic positioning. One plot at a fixed position (e.g. middle plot) within each of the 10 major divisions.

Figure 9.4

Census forms for counting breeding seabirds

SEABIRD COLONY REGISTER

Data Sheet

FOR OFFICE USE

Name: ______________________ Year: ______________

Give address on back of sheet if different from Colony Register Form

Colony Name: ______________________

Notes: Use back of sheet

County or District: ______________________

FILL IN HERE

SPECIES			DATES OF COUNTS	ACCURATE COUNT	+ RANGE OF ESTIMATE min.	max.	Unit	Method	Br. Status
Fulmar	022								
Manx shearwater	046								
Storm petrel	052								
Leach's petrel	055								
Gannet	071								
Cormorant	072								
Shag	080								
Arctic skua	567								
Great skua	569								
Black-headed gull	582								
Common gull	590								
Lesser black-back	591								
Herring gull	592								
Great black-back	600								
Kittiwake	602								
Sandwich tern	611								
Roseate tern	614								
Common tern	615								
Arctic tern	616								
Little tern	624								
Guillemot	634								
Razorbill	636								
Black guillemot	638								
Puffin	654								

FILL IN HERE

UNIT

1 = Individual bird on land
2 = Apparently occupied nest
3 = Apparently occupied territory

COUNTING METHOD

1 = From land
2 = From sea
3 = From air
4 = From photo
5 = From land and sea
6 = Other, give details in Notes.

BREEDING STATUS

01 = Bird in habitat
02 = Singing in habitat
03 = Pair in habitat
04 = Territory
05 = Display
06 = Nest site
07 = Anxious parent
08 = Incubation
09 = Nest building
10 = Distraction
11 = Used nest
12 = Fledged young
13 = Occupied nest
14 = Food for young
15 = Nest + eggs
16 = Nest + young

In Britain a standardised form has been produced for the purpose of counting breeding seabirds (from Seabird Group/NCC 1988; Lloyd et al. 1991).

Numbers of birds counted and estimated are entered in separate columns, so that the two parts of the census when added together give an approximate total for the colony. The following codes define the counting units.

Figure 9.4

(continued)

1 = Individual birds on land, excluding any on non-breeding ledges or loafing areas; 2 = apparently occupied nest-sites; 3 = apparently occupied breeding territories; 4 = other, give details in notes.

The following codes define the counting methods.

1 = From land; 2 = from sea; 3 = from air; 4 = from photo; 5 = from land and sea; 6 = other, give details in notes.

The following codes define the level of certainty that the species breeds in the colony.

01 = Bird in suitable nesting habitat during the breeding season; 02 = bird singing in suitable nesting habitat during the breeding season e.g. petrels; 03 = pair of birds seen in suitable nesting habitat during the breeding season; 04 = bird seen defending territory, two records at least 1 week apart; 05 = courtship displays recorded; 06 = nest site found; 07 = agitated/anxious parents seen; 08 = bird seen incubating; 09 = bird seen building a nest; 10 = distraction display recorded; 11 = used nest found, e.g. broken eggshells, droppings, food remains, etc.; 12 = fledged young present – not used for species that may have travelled some distance e.g. petrels; 13 = occupied nests, contents unknown; 14 = food seen being brought to young; 15 = nest with eggs found; 16 = nest with chicks found.

As with other census methods, the standardisation of method and data recording is vitally important when counting cliff-breeding colonies. To assist in achieving such standardisation, there are both standardised methods for laying out counting plots (Fig. 9.3), and standard forms for recording census data (Fig. 9.4). It is particularly important to record the year, month, date and phase of the breeding cycle (pre-laying, incubation, chick-rearing) when the counts were made as this will have a considerable effect on the interpretation of the results. Standardisation of data collection enables comparisons with other areas and between years.

1. Direct counts

For many cliff-nesting seabird species it is possible to make a direct count of the numbers of breeding birds at the breeding colony, although this depends on visibility, etc. The methodology for directly counting the breeding species is quite similar and some examples are given below.

For the Kittiwake, the breeding population is counted as 'Apparently Occupied Nests' (AON). These are defined as a substantial or well constructed nest capable of holding two or three eggs and occupied by at least one bird on or within touching distance of the nest (Seabird Group/NCC 1988). It is usually possible to recognise an occupied nest by the covering of white faeces, although there are difficulties in dense colonies where the nests can appear haphazardly positioned, and because not all nests will have eggs laid in them, and immature birds often occupy unattended or abandoned nests (Walsh *et al.* 1995). Apparently Occupied Nests should be counted during the late incubation to early nestling period (Nettleship

1976; Heubeck *et al.* 1986; Harris 1987). When a whole colony is being counted, the count is made easier if the cliff is divided into sections and these are counted separately (photographs or rough sketches are helpful to avoid double-counting).

One problem with counts obtained using this method is that some non-breeding birds build nests but do not lay. These nests are generally less well built and less obvious than active nests. If possible they should be excluded from counts; also guano-stained loafing sites should be identified and similarly excluded. The final counts should ideally be the mean of at least three counts of the same section of the colony on the same day. Repeat counts of particularly dense colonies on different days are a valuable way of checking the results, especially if the same observer undertakes all these counts. However, for colonies in northern Europe and where time is short, a single count of the site in June provides a good estimation of numbers as the variation in colony attendance at this time is low. This optimal period will vary with latitude, being later further to the north.

For the Fulmar the breeding population is counted as 'Apparently Occupied Sites' (AOS), defined as an individual sitting tightly on a reasonable horizontal area large enough to hold an egg (Nettleship 1976). Two birds on a site, apparently paired, also count as one site. It is difficult to determine the number of breeding pairs because prospecting birds may occupy a site for several years before producing an egg and may be confused with breeding pairs, because sites are occupied temporarily by birds which do not breed, and because of the relatively extended breeding season (Walsh *et al.* 1995).

Counts of Apparently Occupied Sites should be made in the late incubation to early nesting period when the colony attendance is greatest, usually in late June or early July in northern Europe (Dunnet *et al.* 1979). Normally a number of counts are made over a period of 5–7 days, and a mean value taken, to reduce the influence of variations of colony attendance between days. In northern Europe, counts should also be made in the middle of the day (12.00–13.00 hours BST, maximum 09.00–16.00 BST) as attendance is highest at that time. However, in Arctic and Antarctic Regions the continuous daylight during the breeding season may alter this pattern.

2. Corrected direct counts

For some of the species breeding on cliffs there are problems with directly counting the birds because it is very difficult to determine the relationship between the number of birds seen and the actual numbers of breeding pairs (e.g. Harris 1988). This is because colony attendance varies markedly over the day and season and there are no obvious nests. These problems are particularly severe in the auks, for example in Common Guillemot in northern Europe, and this species is used as an example of a method which has been developed to partially overcome the problems.

In Common Guillemots the breeding population is assessed by first counting individual birds at the breeding colony (Birkhead and Nettleship 1980; Evans 1980). Individuals are best counted during the middle egg-laying to middle chick-rearing period, as colony attendance is most stable during that period (Gochfield 1980; Hatch and Hatch 1989; Fig. 9.5). As counts vary with the time of day, counting is recommended between 10.00–13.00 hours BST in Britain. However, other studies have noted a different diurnal attendance pattern (Evans 1986; del Nevo 1990) so it may be necessary to assess attendance patterns of the target colony before counting is started. Counts should be made on 5–10 separate days over the counting period (Fig. 9.5) and results then averaged to even out between-day differences in attendance. If possible the number of birds nesting in crevices and cracks should also be estimated. Counts should also be avoided during winds stronger than Beaufort Force 4, or during heavy rain and fog. Owing to high variation in attendance at the colony, only large population changes can be reliably detected.

Because of the difficulty of assessing the number of breeding pairs from the counts of individuals, correction factors have to be calculated to convert the counts of individual birds into estimates of breeding pairs. A recommended method to calculate conversion factors ('k') is to define 'control' ledges where the number of birds on nests with eggs (Np) and the mean number of individuals (Ni) are carefully assessed from 10 separate counts during the census period, i.e. between the last egg being laid and the first chick hatched. It is important that the number of birds in these control ledges are counted as accurately as possible (never estimated or counted in blocks of 5 or 10). The conversion factor between counts made and the actual number of birds breeding for this particular control ledge is then calculated as $k = Np/Ni$ (Birkhead and Nettleship 1980).

In order to obtain an estimate of the total number of breeding pairs in a colony, counts of individuals in the colony should be multiplied by the mean k value obtained from all the control ledges. However, it is essential that both the original count data and the corrected counts are presented in any publication. Moreover, the k values may be colony specific, and hence may need to be calculated again for new study areas.

Despite being a successful method for Common Guillemot, this approach does not work for even some closely related species, such as Razorbill and Black Guillemot (because many of the birds breed in caves or rocky screes). Details of methods which can be used for Razorbill are found elsewhere (Ewins 1985; Harris 1989; Bibby et al. 1992; Walsh et al. 1995), and for the Black Guillemot, below.

3. Counts from photographs

Some species of seabirds build either particularly large and obvious nests (often regularly spaced), or are themselves large and obvious. For such species it may be feasible to take photographs of the breeding colonies (from nearby land, a boat or the air) and then count the birds/nests from the photographs (or from a projection of the photographs onto a screen or a wall).

The Northern Gannet is an example of a species which can be counted using photographs, which in this case is possible because the Apparently Occupied Nests are very regularly spaced (Harris and Lloyd 1977). The recommended counting method is to take photographic transparencies of the colony and then project them onto a screen and block out individual Apparently Occupied Nests as they are counted. Tests of observers counting birds from photographs indicate that observer error is usually less than 15%, and with experienced observers can be under 10% (Murray and Wanless 1986). The regular nest spacing means that the population size of the colony can be assessed from sample counts combined with a measure of the colony area; expanding colonies always increase in area, and declining ones always shrink in area.

Common Guillemots have also been counted from photographs of the colony. However, such counts are generally unreliable because the birds are often not obvious on photographs: when viewed from the front they blend in with guano-covered ledges and from the back with rock and shadows. An additional problem is that the number of non-breeders cannot be deduced in this way (Birkhead and Nettleship 1980), but for large colonies that cannot be counted from land, photographs taken from a boat may be a better alternative than trying to make direct counts from the same boat as it rocks in the waves.

4. Counts of birds on the sea

For some of the species nesting on cliffs the population cannot be counted on the cliff as the birds breed in locations that are not possible to see. An example of such a species is the Black Guillemot. Here counts of the breeding population are difficult because nests are generally out of view in cracks and gullies and the species normally nests at a low density. Instead, the breeding population is assessed by counting adult-plumaged individuals on the sea. The recommended counting method is to walk along the top of rocky shores and low cliffs, or drive a boat along the base of cliffs in the early morning (05.00–09.00 hours BST) in the pre-breeding period (April to early May in Britain), and attempt to flush all birds out into the sea. All adult-plumaged birds on the sea within 200–300 m of the shore should then be counted, and all immature-plumaged birds noted separately. These counts are best repeated 3–5 times on separate days to produce an average count, and preferably should be made only when there are calm sea conditions and winds less than Beaufort Force 4 as higher winds make counting extremely difficult and colony attendance is altered. Counting in the early morning is important as the birds fly out to sea later on in the day (Birkhead and Nettleship 1980; Ewins 1985).

Counting methods for ground-nesting seabirds

Many species of seabirds nest on the ground in breeding colonies, which in some cases can comprise thousands of birds. These colonies are generally located in places inaccessible to predators and humans, often small and remote oceanic islands or remote beaches. For many species the location of the colony can be stable from one year to another, but for others the species move their breeding colonies to places where there are suitable nesting conditions in a particular year. Nesting terns in particular often move their breeding colonies between years, hence it is extremely important to search the study area thoroughly to determine the distribution of colonies before the counting commences. There are also problems with some long-lived species taking many years to reach maturity, and where only a proportion of the population breeds in any one year (e.g. albatrosses, frigatebirds).

When counting birds nesting in colonies on the ground it is always important to minimise the time spent in the colony, especially if birds are being counted using transect or quadrat methods. Thirty minutes is a maximum period observers should remain in a colony, and if the birds are disturbed within the first few minutes the observers will have to withdraw until they settle down again. Prolonged disturbance may lead to egg loss (e.g. predation), or chick loss (e.g. wandering from their nests and being either attacked or killed).

A number of methods have been developed to count seabirds breeding in colonies on the ground, and these are described below, with examples where appropriate.

1. Direct counts

In gull (or other seabird) colonies containing fewer than c. 200 pairs, the birds can be counted directly using telescopes or binoculars during the mid-incubation period. In gull colonies counted in this way, the counting units are Apparently Occupied Nests, which are defined as the summed number of occupied and unoccupied nests that appear to have been used during the present breeding season (Fig. 9.5). This caveat is applied because gulls have precocial young and some nests may have fledged their young when others still contain young or have not yet had eggs laid in them. It is also recommended that full nest-counts are made between 09.00 and 16.00 hours as colony attendance is most stable during this period, and that counts are not made during periods of heavy rain, fog or high winds as these are believed to affect the accuracy of the count (Wanless and Harris 1984).

For some other species where the nest site is less obvious and the colony is more diffuse, e.g. skuas, the counting unit is the Apparently Occupied Territory (AOT). In northern Scotland the skua population has been assessed in the mid-incubation to mid-nestling period by plotting all records of birds on a map and interpreting these using methods outlined for Territory Mapping (see Chapter 3). Such an approach allows the number of AOTs in the area to be defined (Everett 1982; Furness 1982; Meek et al. 1983). One problem is that non-breeders may also establish territories, which will inflate the apparent breeding population. Direct counts can also be made of 'clubs' of non-breeding skuas to give an idea of the status of this part of the population (Walsh et al. 1995). Care should be taken to standardise the time of day that such counts are made as the numbers of birds will vary during the day.

Direct counting methods are also used to count numbers of breeding terns in small colonies. Here the counting unit is the Apparently Occupied Nest, defined as those birds sitting tight and apparently incubating eggs or

Figure 9.5

Seasonal
variations in
numbers and
clutches of
gulls'
breeding
colonies

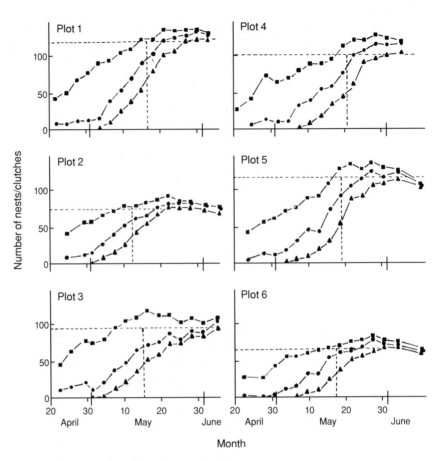

This figure shows changes in the total number of nests (■), complete nests
(including those containing eggs) (●) and clutches (▲) of Larus gulls in six
plots on the Isle of May in Scotland between April and June 1983 (from
Wanless and Harris 1984). The horizontal dashed line indicates the number of
pairs breeding in the plot (taken to be the highest count of clutches plus nests
that had lost clutches during the 12 days prior to that count). The vertical
dashed line shows the median date of laying of the first egg. This figure
indicates that the population of breeding gulls reached a peak in late May and
began to fall in June. Nest-counts should be made during the period of
maximum nests (late May) and counts in future years should be made at the
same time.

brooding chicks. Ideally the Apparently Occupied Nests are counted from a position where the whole of the colony can be viewed. However, problems occur when the whole colony cannot be seen, or where both members of the pair sit slightly apart and both are counted as incubating.

2. Transect counts

In less easily viewed, or larger colonies the number of breeding seabirds, normally counted as Apparently Occupied Nests, may be assessed using transects through the colony. The first stage of this method is to map the extent of the colony either from a ground survey, or more rapidly from aerial photographs, and to mark the boundaries on a base-map. The second stage is to define transects through the colony to obtain a representative sample of the population. A sample of 0.5% of the area of the colony is an approximate minimum sample area, and the number of transects and/or quadrats required to achieve this can be assessed prior to any counting taking place. Transects can be marked with coloured string when they are set out. Each transect is walked by an observer and the number of Apparently Occupied Nests 0.5–1 m either side of the transect line (depending on nest density) is counted and marked by tags or paint (not red and not on the eggs) to avoid double counting. With information on the area of the colony and the area of the transect, the number of breeding pairs in the colony can be calculated (see Chapter 4).

This method is likely to cause very considerable disturbance of the colony, and hence it must be used with great care as predators may steal nestlings and eggs, and the birds may abandon their nests. If there is much agitation by the birds then the observer should withdraw until they have calmed down. In all cases the observer should not remain within the colony for more than 30 minutes without allowing the birds to calm down.

More accurate surveys may be obtained from 'distance' sampling along transects. This includes assuming/estimating the distance from the transect line to the nest (or bird) and then analysing the data using the 'Distance' software package (Buckland *et al.* 1993; Chapter 4).

3. Quadrat counts

It is also possible to use systems of quadrats to sample the population of birds breeding in a colony. Quadrats can be circular and of 300 m^2 (9.77 m long rope fixed to a pole), or larger in areas with lower breeding density. Large quadrats are more easily made square and marked at the corners using stakes. Quadrats can be placed at equal distances along a transect, or located randomly within the colony (Fig. 9.6). A good methodological example of the use of quadrats to count the Black-browed Albatross in the Falkland Islands is provided by Thompson and Rothery (1991). As with transect methods the use of quadrats to sample populations causes considerable disturbance to the breeding birds and hence must be used with extreme care, and the observer must be prepared to withdraw if the birds are seen to be in danger of deserting the colony, or having their eggs or chicks taken by predators.

When the counts have been made, the total number of active nests in a colony is estimated as:

$$\text{total no. active nests} = (\text{mean no. active nests per quadrat}) \times (\text{total area of colony}/\text{area of quadrat})$$

Recounts of the same quadrats in future years allow population estimates to be compared, assuming that the area of the colony is also known and is not dramatically changing. In colonies which are changing their location or extent markedly it may be preferable to re-survey the colony extent each counting-year, and then establish a new set of randomly located quadrats within the colony (Walsh *et al.* 1995).

4. Flushing counts
In some locations, particularly at isolated positions and on small islands when time available for counting is limited, direct counting is not possible and transects and quadrats may be assessed to be too time-consuming and/or likely to cause disturbance to the nesting birds. In such locations flushing counts have proved a useful means to estimate populations of gulls (e.g. Haila and Kuuesla 1982; Hanssen 1982), and terns (Bullock and Gomersall 1981; Whilde 1985; del Nevo 1990). Such flushing counts are regarded as more useful and accurate for terns than for gulls (Walsh *et al.* 1995).

In 'flush counts' the counting procedure is to use a loud noise (e.g. fog horn) to drive all the birds into the air, and thus allow them to be counted. By averaging the results of several counts the mean number of birds at the colony can be estimated. However, there are further aspects to be considered if accurate counts are to result. For example, Bullock and Gomersall (1981) showed by making counts throughout the incubation and post-incubation period, starting in late May (first egg-laying) and continuing until mid-July (first chicks fledging), that the timing of the counts was important. A peak in the number of birds was shown in mid-June (mid-incubation to early nestling) (Fig. 9.5). They also counted birds throughout the day and showed that colony attendance was most stable between 08.00 and 22.00 hours BST, and thus counts should be made between these hours.

Flushing counts of terns have also been related to the true number of nesting pairs by the calculation of a nest-attendance index (Bullock and Gomersall 1981). This index was derived from a small number of 'calibration colonies', where the birds were counted using direct counting and flush counting techniques every 5 days over the breeding season. On the days when the colonies were counted these counts were made at 2-hourly intervals between 08.00 and 22.00 hours BST. Using this technique it was discovered that three flying birds were equivalent to two breeding pairs. The counts from all the other colonies in the area were then corrected using this calibration figure and an estimate of the total breeding population in the whole study area was made. A similar relationship has been demonstrated between flushing counts and direct nest counts at colonies of Common and Roseate Tern in the Azores (Table 9.1).

Table 9.1

Comparison of observed and expected numbers of Common and Roseate Terns in the Azores using a ratio of three flushed birds equalling two breeding pairs (del Nevo unpublished data)

Colony and species	Flush count	Expected pairs	Known pairs	Difference	% Difference
1. Common Tern	70	47	45	2	4.4
2. Common Tern	14	9	8+	1	12.5
3. Common Tern	28	19	18	1	5.6
4. Common Tern	120	80	83	3	3.6
5. Common Tern	190	127	120	7	5.6
6. Common Tern	126	84	85	1	1.2
7. Roseate Tern	32	21	21	0	0

Counts of burrow-nesting seabirds

Some seabirds nest in burrows, typically on islands away from large numbers of ground predators and humans. The burrows used may be taken over from other animals, or may be dug by the birds themselves. Many of the birds nesting in burrows are highly maritime and they come ashore only for the breeding season, often only at night, and are hardly ever seen above ground during the day. These facts make counting burrow-nesting populations particularly difficult (e.g. Hunter *et al.* 1982), especially as it is sometimes hard to ascertain which species is/are breeding in an area.

In general the counting units for these species are Apparently Occupied Burrows (AOBs), defined as a burrow of sufficient depth to support the bird being studied and showing signs of occupation. Occupancy can be determined from the presence of droppings and scrapings in the burrow entrance, a fishy smell in the burrow, discarded fish near/in the burrow, directly with an optical fibrescope, or by playing the calls of the birds being studied down the burrow at night and listening for a response (e.g. Harris and Murray 1981; Harris 1983; James and Robertson 1985; Harris and Rothery 1988). Another method is to place matches or toothpicks upright in the entrances of the burrows and then on a daily basis check if they have been knocked over.

Rabbit burrows are easy to separate from those used by seabirds, but it is more difficult to separate burrows used by different seabird species in mixed colonies. Tape playback of calls and observation of the birds leaving the burrow very early in the morning, or returning to it in the night time may be required to determine the species involved.

Figure 9.6

Techniques
for position-
ing sample
quadrats or
transects in a
colony of
ground-
nesting, or
burrow-
nesting
seabirds
(from Walsh
et al. 1995)

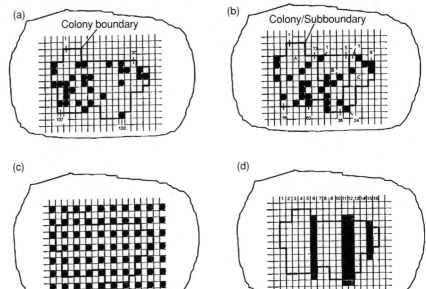

In this simplified example, a preliminary survey has mapped the boundary of a gull colony on the summit of an island. A pattern of grid lines has been overlaid on the map, splitting the colony into 120 grid-squares, each measuring 20 m × 20 m. Nests are to be counted in samples of 25–30 circular quadrats each of area 300 m² (radius 9.77 m), placed centrally within grid squares, or in transects covering a similar area.

a) Unrestricted random (quadrats): 30 quadrats are positioned randomly within the colony as a whole. A series of random numbers is obtained, for example using a calculator. The first 30 numbers (excluding repeats) within the range 1–120 are chosen and the relevant grid squares are marked on the map.

b) Stratified random (quadrats): Three major subdivisions, differing markedly in nest-density have been identified and the high-density (A), medium-density (B) and low-density (C) subdivisions make up c. 50%, 30% and 20% of the colony respectively. Thirty quadrats are positioned randomly within the three colony subdivisions, in proportion to the relative area of the subdivision (i.e. 15 quadrats randomly within A, 9 within B and 6 within C). Squares are numbered separately within each subdivision (A = 60 squares, B = 36 squares and C = 24 squares). Random numbers are obtained as described in (a), except that this is done separately for each subdivision, i.e. 15 numbers chosen for A, 9 for B and 6 for C.

c) Systematic (quadrats): Quadrats are placed at fixed positions within the colony (in this example at intervals of two grid-squares from the approximate centre of the colony). Preferably, the 'starting point' would be detected randomly. Here, 28 quadrats fall within the colony boundary;

Figure 9.6

(continued)

'empty' quadrats which fall outside the colony boundary could be checked rapidly in future years to look for any major increases in the extent of the colony.

d) Unrestricted random (transects): Transects are positioned randomly within the colony as a whole. A series of random numbers is obtained as described in (a) above. The first 3–6 numbers (excluding repeats) within the range 1–16 (the number of potential transects across the long axis of the colony) are chosen, until a maximum of 30 unoccupied grid-squares is included.

1. Sampling methods

Sampling methods are valuable for assessing breeding population of burrow-nesting seabirds. There are several stages to the method, which is similar whatever species are being counted.

Firstly, the limits of the colony are defined on a base map and the species composition is determined. If there is only one species of seabird present then simple sampling methods, using randomly located quadrats or transects (Fig. 9.6), are used to assess the numbers of Apparently Occupied Burrows, and these are used to extrapolate the total colony population. Randomly located circular sampling plots are probably the best method as they are easy to position in rugged topography, and produce a higher number of statistically independent samples than recording burrows along transects. Circular quadrats can be created using ropes tied to a central stake, for example a rope of 1.78 m will produce a circular quadrat of 10 m², 2.52 m of rope will produce a quadrat of 20 m², and 3.0 m of rope a quadrat of 30 m². Smaller quadrats should be used in high density colonies and larger quadrats at increasingly lower density sites. Quadrats should be randomly positioned throughout the colony, and random numbers can be used to define the positions of the quadrats on a map which are then established in the field. By knowing the area of the sampling quadrats and the area of the colony a calculation of the total number of birds in the colony can be made (Wormell 1976; Harris and Rothery 1988). Transect methods (Harris and Murray 1981) and systematic positioning of quadrats (Anker-Nilssen and Røstad 1993) are also used in some places.

2. Mark–recapture methods

It may be possible to estimate the population size of some species by using mark–recapture methods (see Chapter 6). For example, at Skomer Island in Britain attempts have also been made to assess Manx Shearwater populations using capture–recapture ringing methods (Alexander and Perrins 1980). Moreover, on Svalbard Little Auks have been colour-dyed within sample plots. Repeated counts of the numbers of birds which are colour-dyed and those which are not, within the sample plots, allows the total population to be estimated (Isaksen and Bakken 1995).

Monitoring populations

Monitoring animal and plant populations has received considerable
attention during recent years, particularly as a means to demonstrate conservation problems and assess whether recovery programmes are working
or not. Several text-books on the subject have been produced in the UK
(e.g. Goldsmith 1991; Spellerberg 1991; Furness and Greenwood 1993;
Gilbert *et al.* 1998), and similar work has also been undertaken in other
countries (e.g. Engstrom 1990 for USA; Dunn *et al.* 1995 for Canada; Davis
1984 for Australia; Pomeroy 1992 for East Africa).

It is also possible to monitor populations of breeding seabirds in their
colonies. In the British Isles, for example, there has been much progress
over the past 15 years towards developing a unified seabird monitoring
programme (Stowe 1982; Evans 1986; Mudge 1988; Harris 1989; Lloyd *et al.*
1991; Walsh *et al.* 1995; Gilbert *et al.* 1998).

In Britain, the monitoring plots are well defined areas, usually a
colony or a group of birds within a colony, where annual counts of the
breeding seabirds take place. As a general rule, and depending on the
species being studied, these monitoring sections should include 50–100
pairs of cliff-nesting seabirds. The position of the monitoring site within
the colony is important. The plots should aim to provide a representative
sample of the colony. Ideally plots would be randomly located throughout the colony, but in practice randomly located plots may be impossible
to count. At the present time most monitoring plots have been pragmat-
ically selected for their ease of counting and believed representativeness,
i.e. plots encompass most of the variation in the colony, including some
edge, but they avoid areas where birds are particularly densely packed,
where they are extremely difficult to count. Detail of the methodology
used for monitoring numbers of Common Guillemot is presented as
Fig. 9.7.

2. Counting flocking and migrating birds

Many species of birds form flocks for roosting, feeding, migration and protection. The flocks and passages of migrants are impossible to count using
standard methods (Chapters 3–5) and specialised counting methods have
been developed.

A) Flocks of waders, wildfowl and other waterbirds

Although a variety of species congregate in large flocks, populations of
non-breeding waders and wildfowl have received most attention, particularly at the seashore. For example, in Britain waders and wildfowl are
counted within a collaborative scheme involving four organisations and
over 3000 volunteers, named the 'Wetland Bird Survey' (WeBS), which has
developed from many other survey schemes (Owen *et al.* 1980; Prater 1981;
Kirby 1987, 1990; Salmon 1989; Cranswick *et al.* 1996). Similar programmes

Figure 9.7
_____

Procedure for
monitoring
cliff-nesting
seabird pop-
ulations

Details of the methodology used to establish a monitoring plot for Common Guillemot is presented below (after Harris 1989).

Select several study plots (e.g. five) dispersed through the colony where there are 50–100 nests which can be viewed from the same level or from above. Ideally these would be randomly distributed throughout the colony but observer safety and difficulties of viewing in most colonies dictate that at most sites their location will be chosen non-randomly. This can be done by dividing a colony into four or five approximately equally sized sections and picking one or two plots within each section, trying not to bias plots towards the centre or edge of the colony.

Take photographs of the monitoring plots from a good vantage position when the birds are incubating or brooding small young (June in Britain). Large-scale photographs (20 × 20 cm) are essential for the first year, but in subsequent years the outline of the colony, important features and location of study plots can be traced from the original photograph. Tape overlays onto the original photograph so it can be annotated in the field.

View the area from where the photographs were taken, at approximately the same time of year. Plot the positions of (1) birds with eggs, (2) birds with a chick, (3) birds apparently incubating, (4) pairs regularly attending a site that appears capable of supporting an egg (bearing in mind that some eggs are laid on unsuitable sites).

Make several visits until satisfied that most of the occupied sites have been located. Record any chicks without an adult in attendance. Number the active sites. To assess breeding success the contents of active sites should be noted every 1–2 days. Any young leaving when aged 15 days or more and/or are well feathered can be considered as having been raised successfully.

If assessing breeding success, present the results as **x** young fledged from **y** active (i.e. 1–3 above) and **z** inactive (i.e. 4 above) sites as found on the dates of the first checks.

Make notes if you have any reason to suspect the season, or the results, may have been atypical.

Follow the same areas each year.

Similar methods can be used to monitor populations and breeding success of other species of cliff-nesting seabirds.

for counting waders, wildfowl and other waterbirds exist elsewhere in Europe (e.g. Rose and Taylor 1993; Meltofte *et al.* 1994), in the Americas (e.g. Howe *et al.* 1989; Howe 1990; Dunn *et al.* 1995), Africa (e.g. Dodman and Taylor 1995), Asia and Australia (e.g. Davis 1984). Taken together, these programmes, now coordinated internationally by the non-governmental organisation, Wetlands International, are approaching a global coverage allowing the global populations of waterbirds to be estimated and monitored (e.g. Rose and Scott 1994), although for many areas the coverage is still patchy and incomplete.

The regular counts of waders and wildfowl at particular sites, which are now part of national biodiversity monitoring plans in several countries, form a basic part of many scientific research programmes. From them it is possible to derive imprecise population trend estimates for coastal species (Howe *et al.* 1989), produce Atlases showing generalised population density (e.g. Durinck *et al.* 1994; Skov *et al.* 1995), assess conservation needs (e.g. Tucker and Heath 1994), assess ecological and management needs (e.g. Piersma and Ntiamoa-Baidu 1995), and monitor general environmental condition of an area (e.g. Furness and Greenwood 1993). These matters are almost all related to populations at the coast, species which use ephemeral wetlands inland are much more difficult to assess.

The procedures used to count populations of waders and wildfowl is broadly the same for all studies, and is described below. These methods can also be applied to counting other types of birds which are found in large flocks, e.g. migrants, feeding flocks, flocks of roosting birds. The methods of counting larger soaring bird species on migration are presented in Chapter 8.

1. Initial stages

Using maps
In unexplored areas standard topographical maps and Marine Navigational Charts are the best way to find potentially suitable places for flocking waders and wildfowl, although aerial photographs and satellite photographs can be helpful. In areas that are well known and counted every year maps are also used to define the counting sectors, which generally remain the same over time (e.g. Fig. 9.1).

For smaller sites it is desirable to count the whole site from a single viewing position to minimise disturbance and the possibility of birds being counted twice. However, this is not possible on larger sites which have to be divided into counting sectors. The boundaries of the sectors will depend on the habitat being surveyed, e.g. for counts of roosting waders on the coastline, the sector will be a length of coast that can be covered by a counter within 2 hours of a high spring tide, preferably with obvious landmarks at either end. Once defined these sectors form the counting units for future years.

Using tide timetables

When planning surveys of new coastal sites it is essential that tide timetables are consulted in order to assess whether the site can be counted from foot, or whether boats will have to be used. For inland wetlands the size of the wetland and the ease of access will determine whether foot, car, boat or aerial methods need to be used. For many large sites researchers cannot hope to cover the whole site and some areas will have to be selected, often based on the simple criteria of being able to access them safely.

2. Survey methods

Three main methods are used to count non-breeding waders and waterfowl in an area. All have different advantages and disadvantages of coverage, accuracy, access and logistics. Much of the below is taken from Howes and Bakewell (1989), with some material from Bibby *et al.* (1992).

Ground surveys

These are the most widely used and cheap way to gather information, particularly in areas with good road access to different counting sites. However, there are disadvantages in that only small areas can be covered during one day, access on foot is not possible to some locations, and the birds may be disturbed if they are approached on foot.

Two main techniques are used to view a wetland on foot and count any birds present:

1. For lake and coastal shores the observer can walk along the edge of the area, using binoculars to see where concentrations of waterbirds occur. The observer can then stop and, by using binoculars and a telescope mounted on a tripod, count and identify the birds. If only a small number of birds are present the observer can walk and stop every 100–200 m to count and identify the birds. When walking along care must be taken not to disturb the birds otherwise they will fly somewhere else and may be 'double-counted'.
2. Alternatively, a concealed or raised position can be found from which the wetland and the birds can be viewed without disturbance. Positions such as a hill, a tower or a hide are all suitable. Binoculars can be used to locate the birds, and a telescope mounted on a tripod can be used to identify and count them.

Counts are best avoided when the weather is windy or when there is heavy rain. However, counts can be made in much worse weather than for the next two methods.

Boat surveys

Boats are useful for surveying large wetlands as these areas are often inaccessible by foot and aerial surveys are too costly. Boats are generally available locally at a site and can be hired (but not cheaply) from local fishermen, or pleasure boat owners. They permit access to difficult places and can quietly approach birds without causing them to fly away. However, although boats are useful to get around, it should be noted that the identification and counting of birds from a boat is usually quite difficult. Boats are also of no use in bad weather conditions, and as they move more slowly than birds there is a risk of double counting birds that fly away from the boat and land ahead of it.

Survey data can be collected in a similar way to that in other methods, i.e. the survey map can be divided into sections based on features such as hills or rivers, and the birds counted within these sections. It is possible to use binoculars on a boat, but not telescopes. Where possible, it may be better to ground the boat periodically so that the birds can be counted and identified using the telescope (Fig. 9.8).

Boats are also very useful for counting birds on the sea (e.g. Durinck *et al.* 1994; Skov *et al.* 1995), and an outline of the methodology is presented in Chapter 4 (Line transects).

Aerial surveys

Aerial surveys have been used worldwide to identify key sites for waterfowl, to evaluate bird numbers, to map vegetation and identify threats to habitats (e.g. Howes and Bakewell 1989; Durinck *et al.* 1994; Skov *et al.* 1995). They are especially useful in large and remote areas where access to a wetland area is very difficult and study by boat or on foot would be prohibitively time consuming and logistically complicated. In such areas a plane survey can provide an assessment of bird populations in a few hours. The best types of planes are 'high wing' for good visibility, and which have the capability for slow flying and great manoeuvrability. Helicopters have also been used successfully. Two people normally make up the counting team in the plane, a navigator/data recorder and an observer/data recorder. All must have a strong head and stomach as many people become sick during aerial counting exercises. As in other techniques the area to be surveyed should be marked on a map and divided into sections which are counted systematically. Ideally the plane should fly at between 50–100 m; overcast but bright days are ideal for accurate counting (or with the sun from the back) and the best time is during the middle part of the day when long shadows are less of a problem. Fog, rain, haze, thunderstorms, snow and dust storms should be avoided! Observers should wear dark clothes as lighter ones are reflected in the windows which makes the work more difficult. The counting is also better undertaken at high tide when the birds are concentrated in smaller areas and hence are more easily counted. Data are normally recorded in hand-held tape recorders, in note-books and onto maps directly.

Figure 9.8

Boat survey
techniques
(from Howes
and Bakewell
1989)

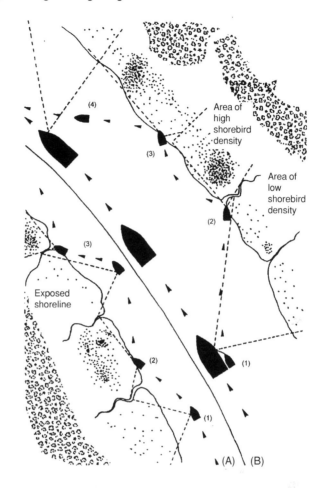

Area of
high
shorebird
density

Area of
low
shorebird
density

Exposed
shoreline

(A): Using a shallow-draught motor boat in shallow water, close to the shoreline:

1. Observe and count low densities of birds from the boat (try to use land-marks, such as channels, to mark the count areas).
2. For large concentrations, ground the boat and count, using a telescope and tripod, standing in water, mud, etc.
3. Move along the shoreline, repeating the necessary technique (NB: avoid disturbance of flocks).

(B): Using a larger vessel with small motor boat in tow, in deeper, offshore waters:

1. Observe and count low concentrations of birds whilst on the vessel, and locate high concentrations of birds.
2. Use the small boat to get closer to these birds by grounding on the shoreline and counting using telescope/tripod.
3. Continue along the shoreline, counting flocks, and,
4. Rejoin the larger vessel until further large concentrations are found.

Aerial counting has the advantage of collecting a lot of data in a short time, and can be used in remote areas which cannot be reached by other means. However, renting the aeroplanes is expensive, and it may be difficult to identify the birds and counts will be best for populations of larger waterfowl, seabirds and waders. Smaller species will either be missed totally, or lumped together.

3. Counting flocks

There are two main methods, and their choice depends on the size of the flock and the ease of counting.

Direct counting

This method can be used in ground and boat surveys. If the congregation is no more than 3000 birds a suitable vantage point should be located and all the individual birds counted directly using binoculars or a telescope. This is easy with large birds close up, but becomes progressively more difficult with large numbers, smaller birds and greater distance. For this method it is important that:

- there are small numbers of birds (less than 3000) present, limited movement of the birds, little disturbance, the birds are in a small area, or they are found in a large open area;
- the observer is a sufficient distance from the birds so as not to disturb them;
- the observer avoids walking directly towards the flock of birds as this might make them fly;
- the observer should try to count the birds with the sun behind, so that the patterns and colours of the different species can be clearly seen;
- the observer should not walk along the skyline as he will be silhouetted and this may scare the birds. It is always useful to try to blend in with the environment;
- the observer should remain quiet and avoid sudden movements as these will scare the birds.

Data collected can be written in note-books, taped on hand-held recorders, or the numbers recorded on a 'tally counter' (a small machine where a button is clicked to count the numbers of birds seen). The tally counter is particularly useful for accurate direct counts, and can also be used in esti-mation counts. As the number of birds in the area increases, and especially if they take to the air, estimation methods have to be used to count their populations (Figs 9.9 & 9.10).

Figure 9.9

Methods of estimating numbers of birds in flocks (from Bibby *et al.* 1992)

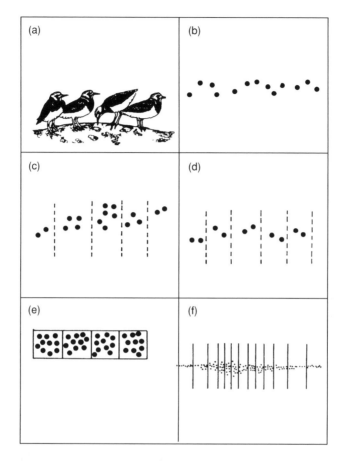

(a) In small roosts and feeding flocks, the number of birds can be counted directly.

(b) For small flying flocks of even density, the birds can be counted individually (1, 2, 3, 4, 5, etc.) to produce an accurate total. If a suitable landmark is present it can be used to help to count the birds.

(c) In unevenly distributed flocks with small groups of varying size, each group of birds should be rapidly counted and added together.

(d) For larger numbers of birds in evenly distributed flocks the birds should be counted in multiples e.g. 2, 4, 6, 8, or 3, 6, 9, 12, etc. Again if landmarks are present they can be used to help divide the flocks in order to count them more accurately.

(e) For densely packed flocks in flight or at a roost, the birds should be counted in estimated blocks. The size of the blocks used (10, 100, 1000, etc.) varies according to the size of the flock. The largest flocks of 10 000 birds or more present the biggest counting problems with even the block method giving a very rough estimate of numbers.

(f) Flying flocks often bunch in the centre. In this case it is important that the blocks are closer together in the centre of the flock than towards the edges, but in practice this may be difficult to achieve.

Figure 9.10

Trends in the
accuracy of
different
observers at
counting
flocks

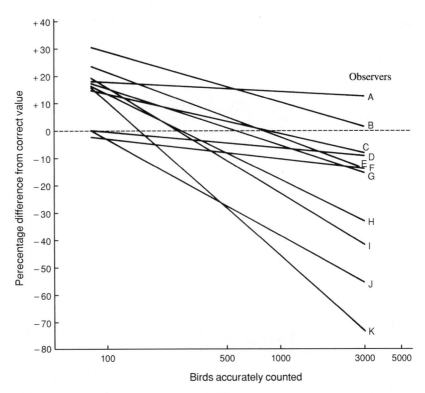

Birds accurately counted

This figure shows that overall there was a tendency for observers A–K to overestimate slightly (10–20%) the number of birds in smaller flocks (100–400),
but to under-estimate increasingly numbers in larger flocks. For example, in
flocks of 3000 birds estimates were consistently low by around 25%. However,
individual observers vary in their ability to estimate numbers of birds. For
example, observer A consistently produces estimates between 10 and 20%
above the true number, but the accuracy of the estimate hardly varies with
increasing numbers of birds. In comparison, observer K estimates 10–20%
more birds in flocks of 100 individuals, but this rapidly changes to an estimate
around 70% below the true figure when counting flocks of 3000 individuals.
This indicates that the accuracy of different observers is highly variable and
needs to be checked in any formal counting programme. (Prater 1979)

Estimation methods

These methods can be used from the ground, boats, or in the air. Estimation procedures should be used if:

- there are large numbers of birds present (more than 3000);
- the birds are continually in flight;
- there is much disturbance;
- the birds are very tightly packed together in a roost site;
- the birds cannot be easily identified due to poor light (e.g. viewing into
  the sun), or the distance between the birds is large.

This method involves counting or estimating a 'block' of birds within a flock, e.g. 5, 10, 20, 50, 100, 500 or 1000 birds depending on the total number of birds in the flock and the size of the birds (Figs 9.9–9.11). The 'block' is then used as a model to measure the remainder of the flock. Landmarks can be used to help break up large flocks into more manageable sections. If possible, counts should be repeated several times and another observer's opinion obtained on the number of birds before a population figure is written down. Problems arise when snap estimates have to be made of a rapidly moving and large flock, and where the flocks are of different density (Fig. 9.11). Counts should be expressed in terms of the species forming the flock (may be mixed), the total number of birds, and an estimation of the numbers of the various species if the flock is mixed, or the proportion of each species. A tally counter is useful for estimation counts and in this case can be used to record the number of 'blocks' quickly and accurately.

4. Counting roosts

Counting shorebirds at their roosting sites is an efficient method of identifying the species composition and overall abundance in an area. In the case of waders at high tide roost sites, counts are made 2 hours either side of high tide on the highest spring tide of the month. These high tide roosting sites are often traditional and it is important that sufficient time is expended to locate them prior to any counting. To locate roosts, all suitable habitats such as saltmarsh, shingle, beaches and spits should be visited on a rising tide when birds are beginning to congregate. Birds may also congregate on short pasture, recently tilled or rolled arable fields or recently harvested fields up to 1 km inland.

Roost counts are made over the period that the birds are present in the area, which may be all year, or for only a short part of the year. The counting area is divided into sectors which are counted regularly, often every month in the case of non-breeding waders and wildfowl on northern European shores. These monthly counts are ideally closely coordinated so that all roosting sites within a single geographical feature (e.g. estuary) are counted as near simultaneously as possible, by a team of counters if the site is large. To obtain sufficient data for comparison of peak counts between years at least five counts should be conducted over the survey period, since it is almost certain that some of them will have to be discarded (because of incomplete coverage, poor visibility, disturbance, etc.). For long-term studies it is also important that there is some continuity of counters and that everyone knows the area well.

In small roosts (a few hundred waders) individual birds can usually be counted from a suitable vantage point at high tide when all the birds are in the roost. Larger roosts, and those comprised of small species, are more difficult to count accurately, and considerable care must be taken when arriving at totals. One, or a combination, of the following techniques is usually successful.

Count all the birds as they fly from their feeding grounds to roost sites, repeating counts where possible. Counts should start at least 2 hours before high water.

Count the stationary birds while they are roosting at high tide, repeating the counts several times. This is the best method as long as the birds are not too tightly packed, as is often the case for small species such as Dunlin and Knot.

Count birds on the ebbing tide when they are leaving the roost, repeating the counts where possible. This method works particularly well for those species that disperse quickly from the roost to start feeding, e.g. Dunlin and Redshank.

At some roosts a combination of all three methods will be needed to produce accurate totals, and with roosts of smaller species which are tightly packed the most reliable estimates will be obtained when the whole flock is in flight and can be counted using methods outlined in Fig. 9.10. Estimates are generally recorded in parentheses, e.g. (3400).

5. Counting foraging birds

When waders and wildfowl are feeding in open intertidal areas, etc., they can also be counted. In most cases the distance between the observer and the feeding birds will be great and the landscape flat. In such instances counting the birds is facilitated by dividing the area, either using natural features, or by positioning canes or poles in the mud at predetermined intervals. These sectors are counted on a cyclical basis with the same pattern of visits being undertaken on each count. All birds feeding and moving within the areas are counted every half-hour. Where possible the observer counts from the first appearance of mud to the lowest part of the tide, or from low water until all the mud is covered.

6. Errors in estimating sizes of flocks

There are a number of difficulties of estimating the numbers of birds in a flock. These are typically because the flocks may contain a very large number of birds, they may have a rapid swirling movement, there may be an interchange of birds between different flocks, the species within a single flock may be of considerably different sizes, some of the birds within the flock may be hidden at any one time, and there may also be problems of poor visibility or with the limitations of binoculars and telescopes.

A few studies have been made on the errors involved in counting flocks of birds (Schuster 1975; Prater 1979; Rapold *et al.* 1985). The study by Prater (1979) attempted to quantify observer error when estimating the size of flocks (Figs. 9.10 and 9.11). Observers were asked to assess the number of birds on a large photograph which had been accurately counted using a binocular microscope. This method found that although individual observers differ in their ability to estimate the number of birds in a flock, the level of error generally varied with the number of birds being estimated. In Prater's study estimates had to be made within 30 seconds, whereas in the field counts may sometimes be made over much longer

Figure 9.11

Problems of
estimating
numbers of
birds in
flocks of
different
density

With large flocks of different density the accuracy of estimations will not be the
same. For example in large flocks which are widely spread it may be possible
to count the birds in blocks, and the large spread of the species tends to give
the impression of large numbers. In flocks which are more closely spread the
estimation method is more difficult to use and the smaller area of the flock may
tend to give the impression of smaller numbers. In fact there are 860 birds in
the upper flock and 1000 birds in the lower flock (Schuster 1975).

periods and repeated several times before a final number is written down,
hence repeated field counts may be more accurate than counts from photographs. However, snap estimates may also have to be made in the field for
large numbers of small species briefly disturbed from a roost; these may be
much less accurate than the counts obtained from photographs. Prater
(1979) has also shown that experience appears to affect the accuracy of
counts produced from photographs of wader flocks. In general the least
experienced counters produced the least accurate results. The detailed
studies of Rapold *et al.* (1985) also indicate large observer errors in the estimation of flying flocks, especially of smaller species. More details can be
found in their paper.

B) Mixed and single-species flocks in tropical forests
Some species of birds in the tropical forests habitually gather into mixed
species flocks. In the South American Amazon this can happen every day
irrespective of the status of breeding in the area (e.g. Munn and Terborgh

1979); in other areas of the world the flocking behaviour varies according to the seasonal changes in the forests. Counting birds in such flocks is a particular challenge, although the distribution of the flocks does not generally overlap in a forest, thus the flock distribution can be mapped, and the numbers of different birds found in each flock can be counted. A crude population assessment can thus be made for these mixed flocks and the species within them.

For single-species flocks such as parrots, which range over a wide area, methods have been developed that try to estimate the average encounter rate with these flocks and the average flock size and species. Over a period of time an estimate of the average flock size and attendance to an area can be calculated as an index of the population. Parrot (and hornbill) populations of rainforest in Indonesia have recently been successfully counted using point count methods (Marsden 1999), and these should give a good estimate of the total popualtion. Furthermore, another recent study has successfully used teams of volunteers to count parrots going to roost at all their known roost sites, which again should produce a good estimate of the total population (Pithon and Dytham 1999).

C) Counting migrants

Migrating birds often occur in flocks, and where these are visible they can be counted using the methods described above for waders, wildfowl and other waterbirds. However, migrants are also found passing one area in small flocks or a few birds at a time over a period of hours/days/weeks. All together these passages may account for many thousands of birds. Specialised methods allow these migrant populations to be counted or estimated as indices which can then be compared between years.

1. Large diurnal migrating species

Migrating raptors (and other large bird species) can be counted at bottlenecks along their migration routes. For example, raptors on migration from Africa to Europe pass overhead in large numbers at The Bosporus (Turkey), Eilat on the Red Sea and at the Straits of Gibraltar, where they are routinely counted (e.g. Porter and Beaman 1985). There are similar key migration points elsewhere in the world, and most of these are now regularly counted (e.g. Fuller and Titus 1990).

As the species concerned are often widely dispersed at other times of the year counting them at their migration points is an efficient way to estimate the population (Meyburg and Chancellor 1989, 1994). Once the methods have been defined, then it is necessary to identify the watch site. Many species can migrate along broad fronts (e.g. Bednarz and Kerlinger 1989), but other species will migrate along leading lines or following topographical features such as mountain ranges or coastlines that are oriented in the preferred direction of travel (Bednarz and Kerlinger 1989; Kerlinger 1989). The methods of migration are thoroughly reviewed in Kerlinger (1989). Once the most likely locations have been located field reconnaissance visits will be required to determine the best watching sites (Bildstein and Zalles 1995). The best sites have at least a 180 degree field of view, and are generally areas of high ground, but there are also considerations of access and safety for the most suitable sites.

Methods for actually counting migrating raptors are summarised in Dunne *et al.* (1984) and Bildstein and Zalles (1995). The birds are best spotted by methodically scanning the sky in the direction from where they are expected. Observers should start at the horizon and scan left to right (or right to left), then move their binoculars up a field of view. This process should be repeated 2–3 times. Between scans observers should look above and to their sides to check for birds that they may have missed. It is helpful to train the eyes to focus at the approximate distance where the birds are first seen as naturally the human eye focuses at 6–7 m distance. Clouds, distant land-marks, aeroplanes, etc. all provide useful reference points for distant focusing. Correct identification of the birds is important and quite difficult, and access to the various flight identification manuals is essential (Ginsból 1984; Clark and Wheeler 1987; Dunne *et al.* 1988). One of the problems of migration counts is that there may be resident birds of the same species in the count area. At most migration points the numbers of birds of different species can be counted directly. These data can be recorded on standard forms, where details of behaviour, weather, etc. can also be entered. When there are large numbers of birds circling in thermals, the best method is to wait for them to start streaming out from the thermals before they are counted.

If the migration flyway is broad, teams of observers (ideally one to three counters, one identification checker and one transcriber) should be positioned to span the flyway, with the distance between them defined by the number of people available, the topography and the need to avoid double counting while accurately counting the birds. Flocks of migrating birds can be counted using methods developed for waders, etc. (see above), or through the use of photographs (Smith 1980, 1985), although the latter method has the disadvantages of being slow, quite expensive and birds can be duplicated on different frames.

It is important that there is training, clear explanations of methods and objectives, proper guidance and standardisation of data collection procedures as these all serve to reduce observer bias. It is recommended that counts are made every day over the migration season as the numbers of birds passing on a single day can vary very markedly. Counts should also be carried out and recorded in hourly units, starting on the same hour to facilitate comparison of results between sites and years. The most complete migration-route counts are made over the entire migration period, about 80–90% of the birds can be recorded over 2–3 week windows, the dates of which are known for most of the important routes.

2. Smaller nocturnally migrating species

Many migrants, especially the smaller species of birds travel at night. Some of these species call to keep in contact with each other. For these species it is possible to assess the numbers of calling birds passing overhead, by using sensitive microphones and special computer programmes (e.g. Evans 1994; Gibbons *et al.* 1996). Currently this method cannot count birds at greater than 1000 m altitude and it also assumes that the birds migrating call at a constant proportion of the whole population, which is not known. However, it does provide a first assessment of the numbers of some species passing certain areas.

It is also possible to count nocturnal migrants through telescopes as they pass in front of the bright disk of the moon (Nisbet 1959; Lowery and Newmann 1966; Alerstam 1990), or to observe them directly using bright lights shone into the air. Migration intensity can be calculated from these counts, and even species compositions to some extent. Radar has also been used to discover migration routes and to calculate the size of migrating flocks (Eastwood 1967; Alerstam 1990; Cooper *et al.* 1991; Bruderer 1992). Radar has also been used more recently to try to assess numbers of Marbled Murrelets in Canada (Burger, 1997).

A recent test of the numbers of birds recorded by moonwatching and radar and infrared devices (Liechti *et al.* 1995) has been undertaken in southern Israel. In this study a pencil radar beam, infrared device and telescopes were all orientated close to each other towards the disk of the moon. Results indicated that about 66% of the birds passing were recorded by people watching the moon, whereas the infrared observations recorded close to 100% of the birds. Birds could also be placed into size classes, but species identifications could not be made.

Summary and points to consider

Counting birds that group together in large numbers is not easy. However, a number of methods have been developed which allow populations to be counted or estimated with a reasonable accuracy. For breeding seabirds and non-breeding waterbirds in particular counting their populations at the sites where they congregate is the best way to monitor their breeding populations and there are many other uses of the data obtained.

1. Breeding seabird colonies
These must be located, described and the birds counted using appropriate methods. Methods vary according to species but are well standardised (e.g. in Walsh *et al.* 1995).
Seabirds can be grouped into those which breed on cliffs, those which breed on the ground (including in trees) and those which breed in burrows. Methods are developed to count populations of species using these various breeding strategies.
Gulls and terns are counted as Apparently Occupied Nests. The counts can be made directly, along transects, within quadrats, or after flushing the birds from their nests.
Auks are counted as individual birds during the peak nesting period. For some species, detailed counts of individuals from small parts of the colony can be used to estimate the numbers of pairs at the colony.
Burrow-nesting species are best counted using randomly located quadrats

or transect lines, but care has to be taken to identify the species of birds in the burrows.

For monitoring purposes, groups of birds in colonies should be counted in sections, or within well defined study plots.

2. Flocking birds

Flocking birds can be counted directly (if the flock is small), or by estimation methods (for large flocks).

Birds in roosting flocks can be counted as they fly in and out of their roost. In smaller roosts, where the birds are visible, birds can be counted directly. Birds in feeding and flying flocks are best counted by dividing a site into sectors before counting the birds. Flocked birds are either counted individually, in small groups, or as blocks, depending on the size of the flock.

The errors in estimating the size of flocks may be considerable and tend to increase with the size of the flock, with an increasing tendency to underestimate the flock size.

For non-breeding waders, wildfowl and other waterbirds, globally coordinated programmes now exist to count their populations and the results of these studies are used for numerous other studies, in addition to monitoring the general environmental health of the world's wetland areas.

3. Migrating birds

Larger diurnal migrating birds can be counted at certain 'bottleneck' sites along their migration routes, using trained observers carefully positioned in the landscape.

Smaller nocturnally migrating species can be counted through recordings of their contact calls, or through observations made against the bright disk of the moon. It is also possible to use radar images to see large migrating bird flocks, and to use the results of long-term ringing studies to assess population levels and changes of nocturnal migrants through time.

10

Distribution Studies

Introduction

A species' distribution can be expressed simply as its presence or absence, or by some measure of abundance, across a of sample units. The sample units may be on a regular grid such as employed in most bird atlas studies, or a random point within a habitat, at which bird data are collected.

There are essentially three types of distribution of animal species and communities: random, regular or aggregated (Fig. 10.1). Birds rarely show a random distribution because this implies they are distributed independently of features on the ground and independently of the presence of other birds. Further, the resources that birds exploit are rarely randomly distributed. For example, songbirds defending breeding territories in a woodland are more likely to be distributed regularly, if within-wood habitat patchiness is taken into account, whereas Sand Martin colonies tend to be aggregated.

The description of the distribution depends on the scale at which the birds are observed which in turn depends on the objectives of the study and the species concerned. Some birds use whole continents during their life-time (e.g. Arctic Tern) whereas others are so sedentary that their whole life may be spent in one particular woodland. The breeding distribution of a territorial woodland bird at the scale of the whole of England, for example, will be aggregated because of the distribution of woods. At the woodland level (or 2×2-km scale) however, they may be distributed more or less regularly because of territorial behaviour, e.g. for Sparrowhawk (Fig. 10.1b). The dispersion patterns of several forest breeding species change as the size of area analysed changes (Wiens 1989). It is important to understand that answers to questions on bird distribution at one scale will not be provided by studying them at a different scale.

A knowledge of the distribution of a species is important because (1) the distribution can be related to land-use, (2) many of the conservation needs of a particular species or community can be identified by investigating habitat preferences which may manifest themselves through patterns of distribution (see Chapter 11), (3) the relative value of sites of conservation importance and vulnerability can be assessed with respect to their bird fauna, (4) information valuable to environmental impact assessments is provided, (5) baseline information is generated against which future

Figure 10.1

(a)

Patterns of
distribution

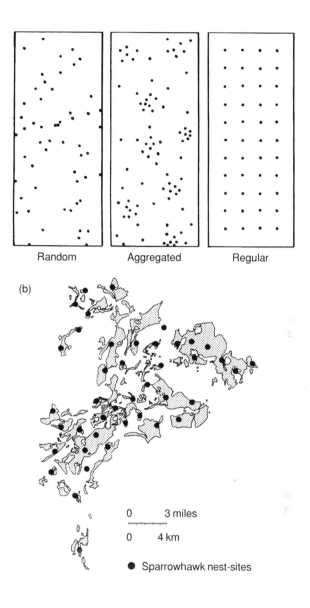

(a) Hypothetical examples of random, aggregated and regular distribution patterns (from Southwood 1978). Birds are rarely randomly distributed because the resources they exploit are rarely so dispersed. Clumped distributions are often observed, for example, colonial nesting seabirds (although even in this example, nests may be distributed regularly within the clump as a result of nest-defence e.g. Northern Gannet).

(b) Regularly distributed nest-sites. Distributions tending towards regularity are largely observed in species that defend a resource, for example, the distribution of nest-sites of the Sparrowhawk shows that they space their nests at a reasonably regular distance from each other at the landscape level (from Newton 1986).

changes can be assessed. Changes to populations may be more obvious from range changes exemplified by a distribution map than from counts (i.e. measures of abundance) taken at the centre of the range.

Distribution studies have been used to identify local, regional, national and international ranges of birds, habitat determinants of bird numbers, the effects of weather, arrival times of migrants, the extent of partial migration, patterns of influx by irruptive species, conservation importance of a particular species, threats to a site and a site's value to conservation.

There are three basic types of distribution study, each of which can be conducted at different scales of detail.

1. Atlas studies: the bird distribution is considered at the international, national, regional or local scale, i.e. on a 'large scale'. Generally atlas studies are presence/absence of bird species, or in some cases abundance measures based on some regular grid-square system across the total area studied (e.g. Birds of North and South America, Britain and Ireland).
2. Single species studies: the bird distribution for a single species is considered at the medium scale, e.g. birds on an island. Look-see methods may be used in which the observer searches for a low-density species based on *a priori* knowledge of its broad habitat requirements.
3. Habitat-based studies: the bird distribution is considered at a small or minute scale by focusing on separate habitats, e.g. birds in a wood, on an estuary, on a heathland. Finest detail of scale can be obtained using radio-telemetry where the individual's ecology and movements are studied.

Atlas studies

The first major atlas of breeding birds was that undertaken by the BTO (Sharrock 1976). Sharrock developed the first system for achieving standardisation, and produced three breeding codes, possible, probable and confirmed, for each species in each 10 km square. An overview of grid-based atlas work is given by Udvardy (1981).

1. Considerations of scale

Atlas studies, for all species present, are conducted at a number of different scales. Obviously the smaller and more fine-grained the scale or grid, the more detail will be attributed to the bird data. Generally four categories of scale exist in bird distribution studies (minute, small e.g. at the scale of a wood, medium and large), and three of these are shown by examples in Fig. 10.2. The largest scale, often used in association with broad habitat definitions, is that derived from the analysis of satellite images, though it is the land cover which is surveyed at this scale, with birds surveyed by conventional field methods being used in association with satellite image data (see for example Pienkowski *et al.* 1990).

Figure 10.2

Different
categories
of spatial
scale in bird
distribution
studies

(a)

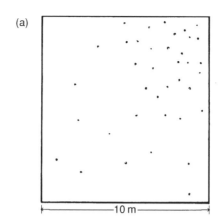

├────────────10 m────────────┤

(b)

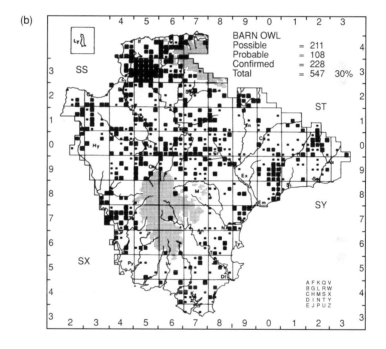

(a) Minute-scale distribution studies. This example shows the position of probe holes of waders in quadrats of 10 × 10 m placed on a muddy shore. A sample of these quadrat units across the shore would reveal patterns of differences in feeding intensity. For an example of a small-scale distribution study at the 5–100 ha scale, see Fig. 10.8.

(b) Medium-scale distribution studies. This example shows the distribution of the Barn Owl, based on tetrads (2 × 2 km squares) from the Devon County Atlas (from Sitters 1988).

Figure 10.2

(continued)

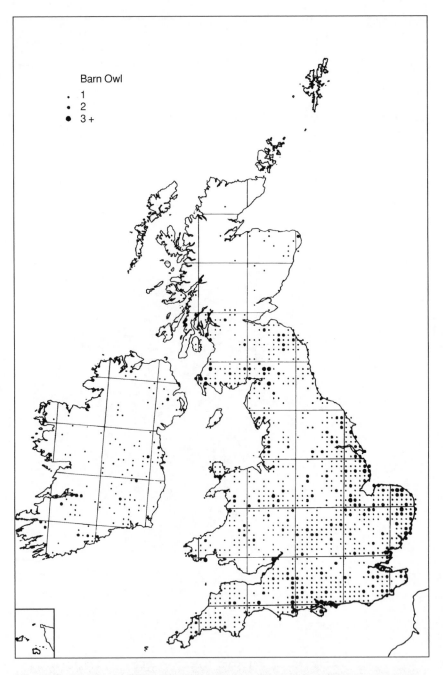

(c) Large-scale distribution studies. This example shows the distribution of the Barn Owl in winter, based on 10 × 10 km squares surveyed as part of the BTO Winter Atlas for the whole of Britain and Ireland (from Lack 1986). In Chapter 1 an example is given of what the Devon Atlas would look like had data been collected at the 10 × 10 km square scale, showing that much detail is lost with respect to the species relationship to land-use and habitat features.

The scale of the study will be determined by the number of field workers available, the detail required, whether the objective is to estimate population size locally, regionally, nationally or internationally (or indeed at all). The capacity to relate to habitat data at the same scale may also be a consideration. The availability of maps is another important factor which is why some countries have used seemingly odd-shaped blocks. National atlasing of bird species distribution has become a major preoccupation of organised ornithology. A hierarchy of working groups, e.g. a national headquarters of an organisation coordinating on-the-ground field teams, is likely to achieve the best coverage and results.

Any atlas is organised on a grid basis. A grid of lines separated by 10 km, 2 km, 1 km, or other scale distance as appropriate and at right angles to each other, divides the area to be atlased into 10 km, 2 km, or 1 km squares, etc. However, countries in which rectangular maps are produced generally adopt these for atlasing, rather than imposing a square grid system.

The following gives the scale of a number of previous bird atlases:

International
    Europe – UTM grid 50 km squares
National
    United Kingdom – 10 km squares
    Netherlands – 5 km squares
    France – Rectangles approximately 23 × 15 km
    Portugal – 20 × 32 km rectangle
    Others in Europe – multiples of 5 or 10 km squares
    Madagascar – ½ degree squares
    Tanzania – ½ degree squares
    Uganda – ½ degree squares
    Kenya – ½ degree squares
    South Africa – ¼ degree squares
    Lesotho – ¼ degree squares
    USA – State-wide e.g. 1 degree blocks, 5 km squares
    Canada – Province-wide, 10 × 10 km or 50 × 50 km
Regional
    Counties in UK – Tetrad (2 × 2 km) or 1 × 1 km in some.

2. Effect of grid size on species diversity

The number of species observed in a grid square increases with observation time and the size of the grid squares, which is directly related to scale. More species are discovered in large grid squares than in small ones, since the former are likely to contain more habitat types. Species diversity can be considered across a range of spatial scales, from continents to variations from point to point within a small copse or woodlot. Wiens (1981), using information from Whittaker (1977), describes seven diversity 'categories' in relation to an increase in area surveyed, or representing a change in diversity across an environmental or climatic gradient or between habitats (Table 10.1).

Table 10.1

Inventory diversities*	Differentiation diversities†
1. For a small or microhabitat sample within a community regarded as homogeneous, subsample or point diversity	2. As change between parts of within-community pattern, pattern diversity
3. For a sample representing a homogeneous community, within-habitat or alpha diversity	4. As change along an environmental-gradient or among different communities of a landscape, between-habitat or beta diversity
5. For landscape or set of samples including more than one type of community, landscape or gamma diversity	6. As change along climatic gradients or geographical areas, delta diversity
7. For a broader geographical area including differing landscapes, regional diversity.	

Levels and types of species diversity

* Inventory diversity refers to that pertaining to a site at various scales of magnitude; essentially derived from a list and/or abundance measures.
† Differentiation diversity refers to that pertaining to a change associated with some gradient.

The relationships between species richness and grid size are non-linear, making it difficult to compare studies that relate to different-sized areas. Further, the number of species per unit area (D) and the proportion of grid squares occupied by a given species (i.e. grid-square frequency (F)) cannot be compared directly when grid sizes are different because the numerical figures change at a non-linear scale. Ellenberg (1985) presents a method of conversion in which plots of density and frequency of species in relation to grid size should be on a semi-logarithmic scale, so that there is a constant number of species added (or subtracted) for each duplication (or halving) of observation effort.

For a range of plot sizes between 10 and 1500 ha the number of bird species breeding in an area and the size of area measured in duplication steps, i.e. 25 ha, 1 km, 4 km, 16 km squares, representing steps of 1, 2, 3, 4, respectively, are strongly correlated (Fig. 10.3). The constants of the linear regressions are characteristic for different broad habitat types.

Further, irrespective of the change of grid size (e.g. 25 ha to 1 km, or 1 km to 4 km), the relative change in F (grid-square frequency) is similar, enabling the calculation of conversion factors (CF) for F as a function of the grid-square frequency in a smaller grid square. Details are given in Ellenberg (1985).

3. Using historical information
Where resources do not permit a full and new survey, historical information is sometimes used. This has been the case for some African atlas

Figure 10.3

Effect of grid size

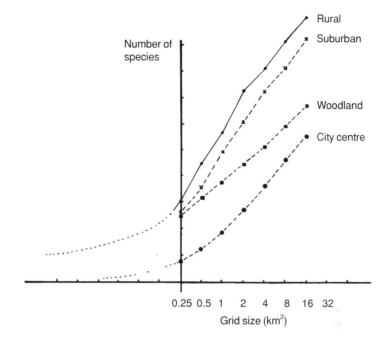

Grid size (km²)

The number of species per unit area and the proportion of grid squares occupied by a given species i.e. grid-square frequency, cannot be compared directly when grid sizes are different because the numerical figures change at a non-linear scale. Ellenberg (1985) presents a method of conversion in which plots of density and frequency of species in relation to grid size should be on a semi-logarithmic scale, so that there is a constant number of species added (or subtracted) for each duplication (or halving) of observation effort. For a range of plot sizes between 10 and 1500 ha the number of bird species breeding in an area (on the y-axis) and the size of area measured in duplication steps (on the x-axis) i.e. 25 ha, 1 km, 4 km, 16 km squares, representing steps of 1, 2, 3, 4, respectively, are strongly correlated. As the plot size within the grid is increased in grid-based distribution studies, more species are encountered. The relationship is almost linear for a range of habitat types varying in structure. The main difference in the lines is due to the higher values of the intercept (on the vertical axis) for samples taken from more structured habitats. This shows that more structured habitats for a given area have more species, based on the assumption that greater structural diversity supports more exploitable niches.

studies, and in part for the European Atlas (Hagemeijer and Blair 1997) for countries with few birdwatchers. In such cases the historical data are in the form of bird reports for previous years, and of county gazetteers which may date, for Britain at least, back to the 18th century. These data have also been used to determine the changes in range of certain species in Britain, notably Buzzard, Capercaillie, Wryneck, Red-backed Shrike, Stone Curlew and Little Ringed Plover. Fig. 10.4 gives an example for the Red-backed Shrike. Usually, however, the assumptions and biases of the data collection

Figure 10.4

Long-term
population
monitoring
using distrib-
ution studies

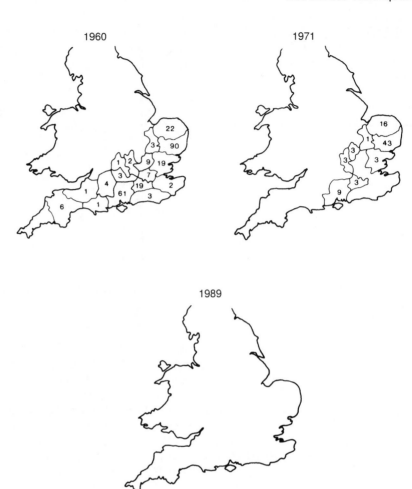

County recording of the distribution of pairs of Red-backed Shrike at irregular
intervals since 1960 shows the decline in the species in southern England
(from Bibby 1973). Here the number of pairs were added up by using county
boundaries as units.

are not known, nor is the sampling method, making comparisons of quantitative with descriptive data difficult to interpret. This approach can,
however, be quite adequate for determining broad changes in range.

4. Planning an atlas
This section describes the planning of an atlas at a national or regional
(e.g. county) scale. Many of the above atlases have endeavoured to use standardised methodology, e.g. the use of the same codes as the European Atlas
for the Ontario Bird Atlas, and the recommendation that atlases in African
countries should use the same scale as each other. The following are
important considerations.

1. The methods must be scientifically valid.
2. The methods must be acceptable in the field to the largely amateur observers.
3. The methods should be the same for all species, although some atlases do employ different methods for different species, particularly common versus rare ones.
4. All data must refer to birds actually recorded. Observers must not be able to send in data based on what they 'know' to be present.
5. The methods must be able to incorporate casual observations otherwise a great deal of potentially usable information will be lost, especially for the rarer and more elusive species.
6. Two or more observers will sometimes be working independently in the same area. The methods must not involve a subjective decision as to which observations to use.

Principles broadly relate to assessing objectives, deciding on scale, developing methods, standardisation, analysis, organisation in the field, computing and producing maps. A diagrammatic representation of the process of designing and conducting an atlas study is given in Fig. 10.5.

Examples of atlases

1. The atlas of wintering birds in Britain and Ireland

This example is given to show the various stages in the process of producing an atlas of bird distribution. A full-scale pilot survey for the Winter Atlas (described in more detail in Lack 1986) was conducted in the winter 1980/81 with two main aims. Firstly, it was necessary to find a method of assessing abundance; secondly, it was necessary to find out about any movements between November and March in order to define the limits of the 'winter', i.e. the months that can be categorised most specifically as winter based on their bird complement. It was decided not to start the field season until the middle of November, and to finish at the end of February before breeding activity commenced in order to reduce effects caused by movements of migrants.

The pilot survey suggested that the number of birds seen on any one day was a good unit of relative abundance. The 'day' was standardised as a period of 6 hours (see below). It was also decided to take the maximum number of birds counted on one day as the measure of abundance. Sometimes this might lead to one particularly large count being used, but this risk was outweighed by major statistical difficulties when calculating means or medians, caused particularly by casual records.

This method has two weaknesses.

1. It is possible to use misleadingly one particularly large count, e.g. a flock.
2. There is only correction for observer effort within a day but not for the total number of days, which may have specific species as well as overall community biases.

Figure 10.5

Designing an
atlas study

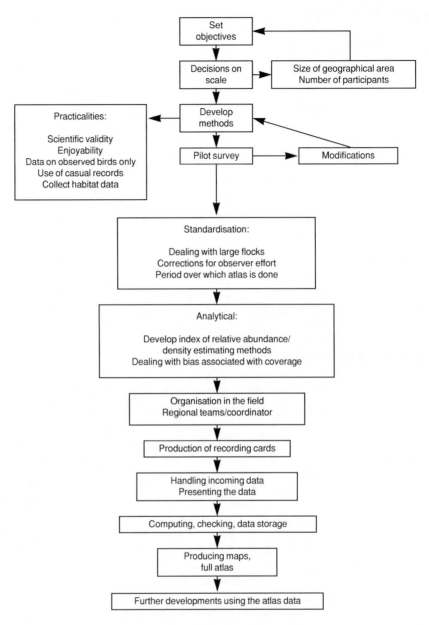

The processes involved in designing and conducting an atlas study are shown. Previously used field methods known to the organiser (**a priori**) can be modified and tested through a pilot survey. One of the most valuable products of an atlas study is often never attempted – that of relating the bird data to habitat, land-use and topographical features, terminating in the development of predictive models which can then be used for conservation evaluation purposes.

The strengths of this method are associated with difficulties when calculating means or medians, particularly so in the case of casual records. For example, a 'zero' count for a species in a restricted habitat, e.g. ducks on a lake, could mean either that no ducks were there or that the habitat was not visited and yet ducks were present.

Two kinds of records were accepted: first, the result of a visit to a 10-km square specifically to do fieldwork for the Winter Atlas; and second, any casual records (termed Supplementary Records) of individual species.

For a specific visit observers were asked to spend a minimum of 1 hour in their 10-km square and to count all birds seen and/or heard. At the end of the visit the total number of each species was recorded on a Visit Card to aid easier computer checking, together with the 10 km square, an identifying feature of the square, the date, the time spent in the field and the total number of species recorded.

A 'day' was defined as 6 hours in the field, because it was considered that this would be the longest that most people would be likely to spend doing fieldwork on a winter's day. In fact only about 3.5% of all cards received were for periods longer than this.

All timed counts of longer or shorter than 6 hours were standardised to 6 hours to permit better comparisons of areas that might have only 1- or 2-hour counts with those that had 6-hour counts. The procedure adopted was to calculate, for each species individually, a coefficient for the regression of number of birds seen on time spent in the field. The data were normalised by putting both axes on a logarithmic scale. With a large number of data points available even quite weak relationships between numbers of birds and time spent in the field are significant at the usual statistical point of $P = 0.05$. As nearly 200 species are considered in the atlas, standardisation corrections were used only if the relation was statistically significant at $P < 0.001$. Many of the commoner land birds came into this category. The majority of the rarer species and those restricted in their habitat preferences have a zero coefficient and no corrections were made. In practice this means you are just as likely to see a rarer or more elusive bird in the first hour of a count as in the sixth or, similarly, visit the restricted habitat, e.g. a lake, in the first or sixth. You are unlikely to accumulate more and more as fieldwork continues. For the commoner land birds though this is what does happen, and therefore the coefficient is positive and standardising corrections are needed by multiplying by $(6/T)^b$ where 6 is the standard 6 hours, T the actual time spent on the count and b the regression coefficient.

The final atlas (as with the first Breeding Atlas by Sharrock) was accompanied by overlay maps of topographical and environmental information, to aid the reader's interpretation of bird distributions.

Problems encountered with the Winter Atlas and their solutions are given in Table 10.2.

2. The atlas of wintering North American birds

This atlas is an analysis of the Christmas Bird Counts conducted since 1900 (Root 1988). Assumptions and refinements to the counts are given in Bock

Table 10.2

Problems
and solutions
in designing
the British
and Irish
winter atlas

Problem	Solution
Roosts	Counts made at roosts should be kept separate from other counts. After a pilot study the difference was not considered important
Habitats crossing 10 km square	Deal with each 10 km square individually e.g. put a lake into a 10 km square in which it predominates
Determining position of square	In the case of estuaries square boundaries were boundaries described by natural points near to the real boundary
Birds flying over the square	Not included in the counts
Uneven coverage explains significant amount of variation in distribution e.g. lowland England received greater coverage than Scottish highlands	The maximum count of each species should be tested for correlation with the number of visit cards received for the square. If there is a correlation then uneven coverage could be biasing the bird distribution
Exaggerated impression of distribution of rare species	Caused by recording presence of one individual found in each of a widespread number of grid squares. Represent rare species independently from the main atlas maps by giving square-specific abundance measure
Flocks or rare individuals moving between squares	Difficult to overcome, apart from using ecological common sense at the regional or local level, e.g. Marsh Harrier recorded in three squares in the Wexford Slobs area in SE Ireland, all probably refer to the same individual

and Root (1981), Drennan (1981) and Arbib (1981). Each count site covers a circle of 15 mile (24 km) radius and at least 8 hours must be spent counting. Twelve hundred or so sites are covered annually on any day within a 2-week period around Christmas. For the purpose of the atlas, mean values counted per site per year over a 10-year period were used to produce computer-generated contour and 3-dimensional maps of distribution and abundance patterns of species in winter. The use of means over a 10-year period is thought to reduce any spurious effects due to weather and abnormal movements of birds. These means summarise the raw data and are therefore an interpretation.

There are a number of deficiencies in the counting procedure. Sites at which counts take place are not uniformly distributed so that there may be biases due to uneven coverage. The abilities of participants, the miles they travel, the hours they spend counting and the size of counting parties, differ between count sites. Further, abundances of gregarious species are inaccurately recorded for two reasons: because flocks are difficult to count accurately; and because chance movements of large flocks can significantly change the recorded abundance of a species.

Variation in count effort at the different sites was diminished by dividing the number of individuals seen at a site by the total number of hours spent counting by the groups of people in separate parties at a site. Mean values at each site were then calculated by summing these values over the various years and dividing by the total number of years the count was held. Since the area counted is restricted to a 15 mile radius, these mean values are densities.

The density values are normalised to range between zero and one for each species by dividing the mean values at each site for a given species by the mean value at the site with maximum abundance. These normalised values are plotted. Maps of species with extremely high (more than 200 individuals counted per hour) or low (fewer than 0.2 individuals counted per hour) abundances are excluded from the main section of the atlas because of difficulties over their interpolation. The atlas also presents overlay maps of elevation, vegetation, mean minimum January temperature, mean winter ocean surface temperature, mean length of frost-free period, mean annual precipitation, general humidity, and national wildlife refuges. Although there are a number of biases and difficulties with the North American Atlas, the vast area covered and the participation required precludes a more statistically valid project being undertaken cost-effectively. This atlas is a good example of the potential of amateur fieldwork.

3. The atlas of the birds of the Netherlands

This atlas was constructed by the Netherlands' ornithological body SOVON (SOVON 1987), from data collected monthly from October 1978 to September 1983 in 5 × 5 km grid squares. Birds both using and flying over the squares were recorded simultaneously, the former often being accompanied by numbers observed. The monthly distribution maps represent a cumulation of 5 years' fieldwork, so that, for example, a map for January contains the results for the Januarys of 1979, 1980, 1981, 1982 and 1983. For most species histograms are used to illustrate their occurrence throughout the 60-month period of fieldwork. The columns of the histograms represent the proportion of squares in which the species was observed in the month concerned, corrected for the number of squares observed per month. Because it has been conducted at the 5 km square level over a 60-month period, this atlas represents one of the most detailed of any country. Overlays of ecological data are also provided at the same scale as that for birds, detailing deciduous forest, coniferous forest, coastal dunes and beach, wet moorland, heathland, marsh, standing water bodies and drift sands.

4. The raptor grid scheme of Finland

The scheme, begun in 1982, aims to collect data on population size and nesting success of Finnish raptors and owls, to establish population trends, and to use nest-site data for conservation purposes. A minimum of 200 hours per raptor grid square (10 × 10 km, chosen freely by a bird-watching group) is spent on fieldwork during February to August. In each square aerial displays of diurnal birds of prey are noted (April), owls are listened for (March), and nests and fledged young are searched for (owls – May and June, respectively, hawks – June and July, respectively). The data are stored in map form thereby constituting a distribution study of both abundance and nesting success, which can then be interpreted geographically.

5. The South African Bird Atlas

This project was established in 1987 and completed in 1993 (Harrison *et al.* in press). Robertson *et al.* (1995) used these data to investigate the relationship between bird atlas reporting rates and absolute density estimates for four species, White-tailed Shrike *Lanioturdus torquatus*, Moneiro's Hornbill *Tockus monteiri*, Ruppell's Parrot *Poicephalus rueppellii* and the Rockrunner *Archaetops pynopygius* in Namibia. Data were collected at the spatial resolution of one quarter-degree square. Reporting rate was calculated as the proportion of the total number of cards for a given square which contain records for the particular species concerned. A sample of quarter-degree squares was subsequently studied in the field using line transects in hillsides, river courses and intermediate areas, recording perpendicular distance from the transect to each bird. Density estimates were calculated using a Fourier series (see Chapter 4) and converted to km^2. This technique did not enable the data to be corrected for time in the field and there may have been systematic bias caused by less effort per card in poorly covered areas. In addition, there was no requirement for a minimum number of cards for each square before the data were acceptable. Despite these problems, the simplicity of the 'proportion of cards' technique makes it an attractive method.

Reasonable correlations were found between reporting rate and the log abundance with up to 28% of the variance explained in some species. Consequently, data from atlases in the form of reporting rates, can be used to estimate abundance. The method is most useful for commoner species; rare or secretive species require more intensive sampling. Bias associated with the atlas data includes inconspicuous species that are most audible during the crepuscular hours being under-represented; prior knowledge of squares may result in observers favouring particular areas or devoting different amounts of effort to different squares. The latter can be remedied by spending a standardised amount of time within grid squares. Bias in estimating population size is introduced where the number of reporting cards is very low.

6. The atlas of breeding birds in Britain and Ireland 1988–1991

This atlas is the successor to the previous atlas of Britain and Ireland undertaken from 1968 to 1972. Observers visited a minimum of 8 tetrads (2 km × 2 km) of their own choice within each 10 km square (Gibbons *et al.* 1993). Two hours were spent in each tetrad during 1 April–31 July over a three-year period. From the timed visits an index of abundance of each species in each square was calculated. Supplementary (non-timed) counts were also collected in order to complete the species lists. If an observer visited a 10 km square that had been visited in a previous year, they were asked to cover tetrads that had not already been surveyed to prevent them visiting their preferred areas.

The frequency of occurrence of each species in each 10 km square was expressed as the proportion of tetrads visited in which each species was recorded. A previous pilot survey showed that these indices were correlated with a measure of absolute density. To overcome the problem of similar frequencies of occurrence for colonial species in which colony size was very variable, observers counted the number of apparently occupied nests of colonial breeders. Counts of coastal seabird colonies added another problem and so data were incorporated from another survey – the Seabird Colony Register. Counts were also required for another group – rare species. The between-squares variation for these species was expected to be low, so observers were asked to count individuals of all species with less than an estimated 10 000 breeding pairs in Britain.

The timed counts were considered vital in order to control survey effort and make the data comparable across the country. Rather than spend more time in a tetrad observers were asked to survey more tetrads thereby improving the precision of the estimate.

A species was considered to be breeding if any of the following activities were observed:

Bird apparently holding a territory

Courtship and display; or anxiety call/agitated behaviour of adult indicating presence of young or nest

Brood patch on trapped bird

Adult visiting probable nest-site

Nest building (including excavating nest hole)

Distraction display or injury feigning

Used nest found

Recently fledged young

Adult carrying faecal sac or food

Adult entering or leaving nest-site in circumstances indicating occupied nest

Nest with eggs found, or bird sitting but not disturbed, or eggshells found near nest

Nest with young, or downy young of ducks, gamebirds, waders and other nidifugous species

The methods of data collection enabled two different maps to be produced for each species; one of abundance and one of distribution, perhaps the first time this has been attempted. The former were based on data collected during timed tetrad visits only, and were thus corrected for variation in fieldwork effort. The latter were based on these and the supplementary non-timed data. One advantage of the data collection at the resolution of the tetrad was that presence/absence data for each tetrad was also yielded. Furthermore, the new atlas methods are easily repeatable so that any change in distribution is more likely to reflect a real effect than if data were collected under less standardised conditions. One disadvantage is that rare and elusive species, particularly nocturnal ones, could be missed during two-hour visits, and subsequently found only by hard searching outside the timed periods. Throughout the data collection a balance had to be struck between counting each tetrad more thoroughly (spending more time in each) and counting more tetrads (thereby also improving distribution information). Finally, methods involving less sophistication (as for example with the Winter Atlas of Britain and Ireland) are less prone to being misunderstood and may be undertaken more enthusiastically than the methods employed in the new breeding atlas; striking the balance between quality of data collected and enjoyment of the fieldwork is therefore important.

Single-species studies

Distribution studies of single species may require a more fine-scale approach than the larger-scale atlas studies of all species described above, but this depends on the original objectives of the study. It is important that the reasons for conducting a single-species survey are identified at the outset and that the methodology maximises the potential for interpretation, for example the calculation of densities and their relation to habitat variables.

The first example given is that of the Fuerteventura Stonechat, one of the most localised bird species occurring in the Western Palearctic and endemic to the arid island of Fuerteventura in the eastern Canaries. This island was visited for 16 man-weeks from 18 February to 11 March 1985 with the aims of estimating the numbers and distribution of the bird, describing its habitats and assessing its likely future welfare (Bibby and Hill 1987).

Fuerteventura, with an area of 1653 km^2, is sparsely vegetated and mountainous but also has stony plains and sand dunes. The survey was based on 21 blocks each of 12 1 × 1 km grid squares (Fig. 10.6). Inland the blocks were 3 × 4 km rectangles. The central squares of the blocks were selected randomly from a list of all squares on the island. If surrounding squares contained no land, an adjoining terrestrial square was selected. The sampling pattern was determined in advance and bore no known systematic relationship to physical features, vegetation or likely suitability for chats. Blocks were visited only once by field teams working at an intensity of about 2–3 man-hours per km^2. Pairs had nests, or, in many cases, fledged young and were generally noisy and conspicuous in open terrain.

For the purpose of analysis, 1 km squares were excluded if they had not been completely covered for lack of time. Some of the random blocks were

Figure 10.6

Sample
positioning
to relate dis-
tribution of a
single
species to
large-scale
habitat
features

Using the 1 km grid map of Fuerteventura in the Canary islands, sample grid squares to be searched for the Fuerteventura Stonechat were determined randomly. Filled squares had breeding season records from 1985,1984 or 1979; open squares were fully covered in 1985 but had no records of breeding chats. Topographical and habitat features of the squares with and without chats could also be recorded and used to relate, without bias, these features to bird abundance (from Bibby and Hill 1987).

less than 12 km^2 in area because they adjoined the sea. The 21 blocks had a land area of 235.2 km^2 of which 209.8 km^2 (89%) was searched. The results are a random sample of 12.7% of the land area of Fuerteventura. A later section of this chapter describes how the information was used to estimate population size.

A special case of single-species studies applies to rare or low-density species using look-see methods. The amount of effort required to sample the species using random sampling would be inappropriate and too costly of time in such a case. Using *a priori* knowledge of the species requirements, the method entails searching the most likely areas in order to maximise the rate of encounter of the species. Habitat measures taken at localities where the species is present can be paired with sites in which the species is absent. (See Chapter 11 for more information.)

Habitat-scale studies

Mapping the distribution of species in specific habitats is done for two main reasons: (1) to relate the distribution of bird registrations derived from territory mapping to the availability of habitat types, thereby enabling preference or avoidance to be determined; and (2) for specific management objectives.

1. Bird registrations and habitat types

The distribution of Nightingales (and other songbirds) in Ham Street Woods, Kent, an area of actively coppiced woodland, shows a preference for 6–7 year coppice stands over older and younger types (Fuller *et al.* 1989). In this study, birds were counted by a single observer using territory mapping (see Chapter 3), over a 5-year period. Each year 23–25 visits were made, spread throughout the breeding season from late March to early July. Each year the entire site was covered evenly and coverage (effort) was consistent between years. One important modification to the methodology was that, because the site, with 52 compartments (blocks of woodland), was extremely complex, it was impossible to assign most of the 'territories' to one compartment or another. Therefore, the densities of 'registrations' recorded in compartments of different ages were used as indices of abundance. The map of the registrations for Nightingales is shown in Chapter 11 (Fig. 11.2).

The age of coppice (number of summers' growth) for all compartments was known and used to investigate the density of registrations in coppice of varying age (Fig. 10.7).

For each year class of coppice the total number of registrations recorded in that year class was divided by the total area. The resulting index was expressed as the number of registrations per hectare. There are two potential problems with interpreting the derived patterns of bird distribution: (1) individual compartment effects (e.g. coppice composition, soil type) could be confounded by year-class effects, but as compartments within each age class were widely distributed this was not too severe; (2) it is

Figure 10.7

Distribution of a single bird species in relation to smaller-scale habitat features

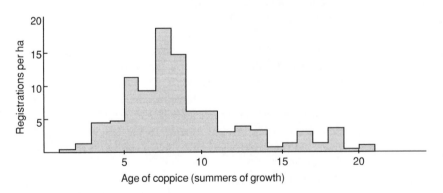

This figure shows per ha registrations of Nightingales in relation to the age of coppice (cyclically cut) woodland. The coppice age categories holding the highest densities of registrations are taken to be those preferred by Nightingales. This shows an example of a distribution study from which data are interpreted in relation to habitat type (from Fuller *et al.* 1989).

not possible to test statistical deviation of the patterns away from randomness because the registrations upon which they are based are not
independent samples but are, in many cases, repeated observations of the
same individual. This could be overcome if results were based on territories.

Roost and feeding sites of waders on estuaries provide another example
of distribution within a habitat. There may be fine-grained environmental
changes across the estuary which drive the distributions of birds. Such plots
reveal the aggregated distribution of waders on some estuaries in which
honey-pot areas hold most of the birds. Fig. 10.8 shows an example for the
Severn estuary.

2. Nest distribution patterns

We have already given an example of the use of maps of nest-site locations
in explaining the distribution of breeding Sparrowhawks (Fig. 10.1b).
Here a further example is given of the use of nest distribution patterns in
relation to vegetation for Mallard and Tufted Duck. Regular searches of a
small island on a lake in Buckinghamshire were conducted to find all duck

Figure 10.8

Aggregated
distribution
of a group of
birds in
relation to
habitat
features

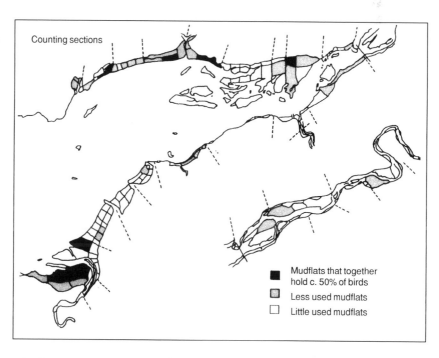

Counting sections

■ Mudflats that together
hold c. 50% of birds

▨ Less used mudflats

☐ Little used mudflats

In this case the whole of the Severn estuary in England was divided into manageable units and wading birds (many species) were counted. The counts for
all units were ranked and those making up a cumulative 50% of the total were
shaded black. The map shows that the majority of birds are using only a small
percentage of the total area of the estuary, which infers that resources also
have an aggregated distribution. Note that, compared with Fig. 10.2, this
example represents a small-scale distribution study (from Clark 1990).

Figure 10.9

Duck nests
in relation to
vegetation
types

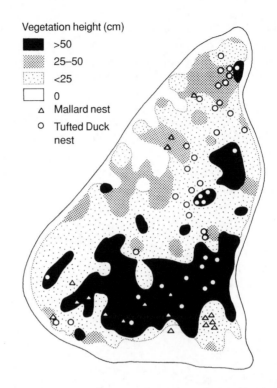

Vegetation height (cm)
- ■ >50
- ▦ 25–50
- ▫ <25
- ☐ 0
- △ Mallard nest
- ○ Tufted Duck nest

In this example vegetation height was measured at each intersection of a grid of 5 × 5 m placed across the island on Willen Lake in Buckinghamshire. Four height categories are represented on the map by different shading, having first interpolated similar heights into contours. A plot of Mallard and Tufted Duck nests is overlaid on the vegetation map. By calculating the area of each vegetation height and relating to the observed and expected (based on area) number of nests, selection for the tallest height category is demonstrated (from Hill 1982, 1984b).

The number of Mallard and Tufted Duck nests is shown in patches of vegetation of different heights. (Chi-square comparing observed with expected nests = 18.8, degrees of freedom = 3, n = 62 nests, P < 0.005). Nests were placed in taller vegetation than would have been expected on the basis of the area of habitat available (from Hill 1982, 1984b). The two species were analysed together since their fate (e.g. predation) operates on the population of nests rather than the nests of the two species independently, since their breeding seasons overlap.

Vegetation height (cm)	Area of vegetation of this height (ha)	Observed nests	Nest density (no. per ha)	Expected nests
>50	0.83	26	31.3	15.3
25–50	1.33	23	17.3	24.5
<25	0.52	12	23.1	9.6
0	0.68	1	1.5	12.5

nests and to monitor their fate. Vegetation height across the island was also measured. The resultant contours of vegetation are shown in Fig. 10.9 overlaid with a map of the nest positions. Observed nest density is compared with that expected on the basis of no selection for height. A chi-squared test was performed (Fig. 10.9) but a more rigorous test would have been that of Neu *et al.* (1974).

3. Radio-telemetry

Distribution data obtained by radio-telemetry are less biased by the observer than those collected as part of a survey count or census, and are particularly useful for determining micro-habitat selection by a species, such as selection of a small group of shrubs or of one crop type in preference to another. Firstly, the radio-locations are marked on a map (or input to a computer database which can generate them in relation to a study area map), and the home range is then determined by joining the outermost points (minimum polygon method) or by some probabilistic method, such as by calculation of the harmonic mean or multi-centred clustering, which identifies core activity areas (see Dixon and Chapman 1980; Kenward 1987). Fig. 10.10 gives an example of the use of radio-telemetry in studying habitat selection by Pheasants (Hill and Robertson 1988). An example of a multi-centred clustering and of a minimum polygon area plot for the same data from the Pheasant study are also shown, to illustrate the different areas produced by the two methods (Fig. 10.10).

The amount of use of a habitat as determined by radio-telemetry, or the numbers of registrations of birds within it, can be related to the availability of that habitat using various preference/avoidance indices (Ivlev 1961; Jacobs 1974). Such indices provide only a ratio of habitat use to habitat availability and do not provide a statistical test. A number of other models use a common statistical approach; Alldredge and Ratti (1986) compare four such techniques (Quade 1979; Neu *et al.* 1974; Iman and Davenport 1980; Johnson 1980). Most use a chi-squared goodness-of-fit analysis to test whether observations of habitat use follow the expected pattern of occurrence based on habitat availability.

The models test slightly different hypotheses and a number of the following assumptions are applicable to different models. (1) All observations are independent in that location 'fixes' are collected far enough apart in time for there to be no temporal serial correlation, i.e. the presence of a bird at position x at time a should not affect its position at time b. Technically, over short periods of time the observations of the same bird cannot be independent. Furthermore, at its worst, numerous radio-locations are really pseudo-replicates. (2) The sample size is sufficiently large to allow a chi-square approximation for the goodness-of-fit statistic (i.e. more than 1 expected observation in each habitat category and less than 20% of all categories contain fewer than 5 expected observations. (3) Habitat availability is the same for all individuals. (4) Results from one animal do not influence results from the other animals. When the number of habitats is small, and with more than 20 animals and 50 observed locations per animal, the Neu method performs best. One

Figure 10.10

Radio-
telemetry
used to
determine
distribution
of individuals

(b)   Radio-locations

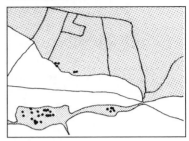

Minimum polygon area

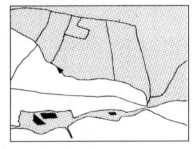

85% multinuclear testing

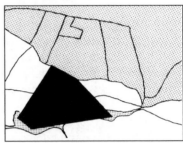

(a)

Radio-location/unit area

+50 50 40 30 20 10 10 20 30 40 40+
Distance inside (m)   Distance outside (m)

(a) Radio-locations to determine habitat preference. Radio-locations of Pheasants, assigned to the distance of the bird from a woodland edge (both inside and outside the wood), show a significant preference for edge habitats. One immediate interpretation would be that the Pheasants, feeding just outside the edge of the wood, move into the edge on the approach of the radio-tracking observer. However, locations were taken at an adequate distance to overcome this problem (from Hill and Robertson 1988).

(b) Radio-locations used to determine home range. In the first example, the minimum polygon area (MPA) method is used to determine the home range of Pheasants; in this method the outermost points are joined together. This gives a totally different value to that when home range is calculated by the multinuclear clustering method, producing a probability contour. The MPA method fills in more non-used habitat than the others (from Robertson et al. 1990). See also Kenward (1987) and Dixon and Chapman (1980) for the harmonic mean contour method.

warning – with all methods, as the number of habitats used increases, the multiple comparison error rates increase; therefore the number of habitats considered should be limited in the study design.

For radio-tracking data, in which visually obscured individuals can be tracked without disturbance and hence bias, the area of each habitat in the core areas (or 85–95% range contours) can be calculated and the observed number of locations in the area compared with that expected if the bird wandered randomly throughout the area defined by its home range contour, polygon, or some pre-determined sample area. The use of compositional analysis (Aitchison 1986; Aebischer and Robertson 1992), overcomes various constraints posed when habitat use is presented as proportions and analysed by chi-square methods. Compositional analysis involves using a 'logratio' transformation on the proportions of different habitat types available and used by the bird. One of the principles underlying the method is that all habitat types should be considered simultaneously because the results from analysing one type at a time are not independent, simply because if one habitat type is used a lot, at least one other will be used less as a consequence.

Examples of the use of distribution studies

1. Estimating population size

Four examples are presented in which population size is estimated from well designed distribution surveys. A fifth example is given where county atlas data have been used to monitor changes in population abundance and distribution over time.

The first is for the Fuerteventura Stonechat, for which the study design has already been described. Total population estimates were made from the number of known pairs and from the total numbers of males (pairs plus extra males). Densities varied considerably between plots, and since each was a different area, variances of the estimated total populations were calculated by jackknife (Miller 1974) thus:

Proven pairs: mean = 591, 95% confidence range = 500–682
All males: mean = 779, 95% confidence range = 663–893

If this survey was 90% efficient and half the extra males found actually represented breeding pairs, a round estimate for the population in 1985 would be 750 ± 100 pairs.

Within the formally randomised census, 196 non-coastal squares were searched and the numbers of males per square were regressed on land-form factors measured from maps. In particular there was a strong correlation between number of males per square kilometre and topography, as measured by the number of 20 m contour intersections of 1 km square boundaries. This model was then used to estimate a total population of 880 males (95% confidence range 792–962) for the whole island based on the whole island's topographical values, and using the model, the

Figure 10.11

Distribution
data used to
calculate
population
levels

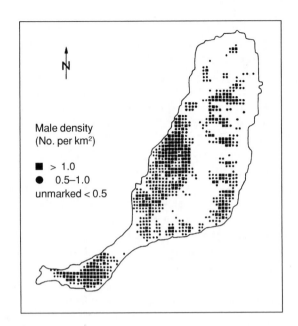

The topography of Fuerteventura was measured by the number of 20 m
contour intersections of the boundaries of each 1 km square, and related to
the number of male chats per square km. The regression line fitted to the
original data is $y = 0.027 + 0.024x$, $r_{194} = 0.342$, $P < 0.001$. This regression was
then used to predict the distribution map of Fuerteventura Stonechats on the
map-derived measurement of topography of the whole island (from Bibby and
Hill 1987).

distribution map for the chat was predicted (Fig. 10.11).

The second example is for the Wood Warbler breeding in Britain (Bibby
1989). Wood Warblers have a very distinctive and carrying song and are
readily detected so there is close coincidence between single visit survey
results and actual numbers (Bibby 1989). These birds therefore lend themselves well to a single-species survey based on fewer visits than is involved in
mapping studies. The objective of the survey was to count a random sample
of one-third of those 10 km squares with proven or probable breeding in
the Atlas of Breeding Birds (Sharrock 1976) in northern and western
regions where full cover was not thought achievable. For the rest of
England it was intended to count all such atlas squares.

Observers were asked to make a single visit to all suitable habitat in each
chosen 10 km square during 15 May to 10 June. Counts were returned as
totals by tetrad, or else tetrads were recorded as unsuitable for these birds.
The total number of singing birds in those 10 km squares in which the
species probably or certainly bred in the period of the atlas was estimated
by assuming that the samples covered were a random selection.

Because of the non-normal distribution of counts and differing sample
fractions between the different regions of Britain, Monte Carlo simulation

Table 10.3

Distribution data used to estimate the British population of Wood Warblers (from Bibby 1989)

Region	Scotland	Lakes	Marches	Wales	South west	Rest of England	Total
10-km squares covered	69	11	21	47	26	243	
Total birds recorded	749	215	290	1636	610	1936	
Mean birds per square	10.9	19.6	13.8	34.8	23.5	8.0	
Total squares in region	317	43	71	159	83	430	
Estimated population	3441	840	980	5535	1947	3426	16 169
Upper confidence limit	4020	1130	1280	6390	2390	4160	17 560
Lower confidence limit	2880	590	730	4720	1550	2860	14 850

(Buckland 1984) was used to calculate confidence limits, by taking the average for each statistic from 10 simulation runs, each of 500 trials. The estimate for the whole of Britain using these techniques was 16 170 singing birds, with fairly wide confidence limits (14 850–17 560) (Table 10.3). The range is high because the number of birds in a square varied widely (from 0 to 253) so it is difficult to be confident about estimating numbers in squares not counted. Further refinement, giving an estimate of 17 200 ± 1370, was achieved by including 'possible' atlas records.

The third example is for the breeding Lapwing survey conducted in Britain in 1987 (Shrubb and Lack 1991). First, every 10 km square in England and Wales that contained some land was identified. Within each a tetrad was randomly chosen, even if the tetrad within the 10 km square fell on water. Observers were asked to visit these chosen tetrads and count the numbers of breeding pairs (males displaying, females incubating, pairs occurring together) of Lapwings by looking into every field in that tetrad. Habitat type was also recorded. One figure for the total number of pairs in the tetrad was produced. The counts were found to be skewed, with 60% of counts having zero Lapwings, if one includes 'counts', for tetrads located in the sea. The sample counts were multiplied by 25 to give the number of pairs in each 10 km square. This was done regionally (nine MAFF regions), and the regional values were summed to give a figure for England and Wales. Because the sample counts were skewed, 95% confidence limits were calculated by bootstrapping methods (Efron 1982).

The fourth example concerns the use of aeroplanes to count 15 major sections and 44 subsections of coast and wetland in Denmark (Komdeur *et al.* 1992), in which both total counts and counts along transects are undertaken, using natural features to divide areas into sections. Count areas are first identified onto maps of a suitable scale (1:200 000). Two observers record on each side of the plane (one observer may be sufficient for special studies of one or two species, especially conspicuous ones). Navigational equipment is required, and it is necessary to decide on the width of transect and the extent which is visible from the plane. With many species widths of 180 m are used; 280 m with seaducks because they are easier to observe and identify. Calibration of the widths at specific heights needs to be done using fixed poles. A total survey is conducted by counting along all shorelines at a distance of 300–400 m. Be aware that aircraft can scare the very species being recorded. Some species are very difficult to identify from above, for example distinguishing between Bewick's Swan and Whooper Swan.

The fast speeds flown by aircraft means that only small flocks can be counted – large ones have to be estimated. All flock locations and numbers are written out and counts summed for each species within each counting area. Fig. 10.12 shows two examples of aeroplane transects in coastal waters. Species distributed along the coastline or which tend to assemble in large flocks are surveyed differently from species which are randomly distributed. The figure also shows the distribution of Scaup assigned to different count sections together with an estimate of numbers.

Advantages of the method of undertaking aerial surveys such as this are quick coverage of large waterbodies, surveys only need a small number of people, it is easy to move between different habitats, and that it is a relatively cheap method per square kilometre surveyed.

Disadvantages of the method are that flock counting is difficult in a fast plane, some flocks remain unidentified, rare or scarce species are overlooked, there is no information on biological or hydrographic factors influencing bird distribution. It is a logistically complex method and time consuming to set up. It is sensitive to weather conditions, largely because of the demands of the aircraft and problems with fog, which can make counting impossible.

The final example is the use of two atlases of Howard County, Maryland, in the United States (Robbins 1990). The total number of records (sum of the species lists) for the 136 quarter-blocks surveyed in 1973–75 was 8297, compared with 9683 for 1983–87, a 16.7% increase. This increase in efficiency of coverage was multiplied by the number of blocks in which a species was recorded during the first atlas period to obtain an estimate of the expected number of blocks for the second period, assuming no change in population. A major departure from this estimate is assumed to indicate an increase or decrease in the local population of the species. Results of this analysis for some species are shown in Table 10.4.

See also Robertson *et al.* (1995) for an analysis of the use of atlas data to estimate population size of some Namibian birds.

2. Relating distribution to environmental data

This is one of the most important uses of distribution studies. Grid-based environmental data are available from a number of sources (though their quality is highly variable), and can be related to bird data at the appropriate

Figure 10.12

Examples of
aeroplane
transects in
coastal
waters

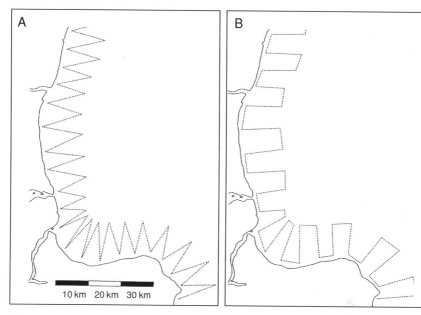

(a)

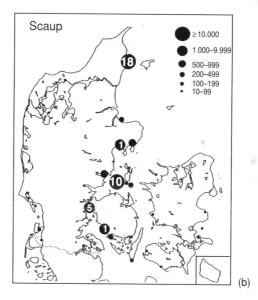

(b)

(a) Two examples to illustrate how aeroplane transects (A,B) may be performed in coastal waters. Species distributed along the coastline and species tending to assemble in large flocks are probably best surveyed by flying in an A-transect; randomly distributed species are probably best surveyed by a B-transect.

(b) Distribution map of Scaup in Danish coastal waters showing the results of the total survey performed in January/February 1989. (From Pihl and Frikke 1992, in Komdeur et al. 1992.)

Table 10.4

Using local
atlas data to
monitor
populations

Species	1973–75	1983–87 Expected	1983–87 Observed	Change (%)
Increases				
Canada Goose (**Branta canadensis**)	8	9	46	+411%
Mallard (**Anas platyrhynchos**)	49	57	86	+51%
Black Vulture (**Coragyps atratus**)	18	21	73	+248%
Pileated Woodpecker (**Dryocopus pileatus**)	29	34	86	+153%
Cliff Swallow (**Hirundo pyrrhonota**)	6	7	20	+186%
Hooded Warbler (**Wilsonia citrina**)	43	50	74	+48%
House Finch (**Carpodacus mexicanus**)	0	0	129	–
Decreases				
American Black Duck (**Anas rubripes**)	10	12	3	–75%
Whip-poor-will (**Caprimulgus vociferus**)	30	35	21	–40%
Horned Lark (**Eremophila alpestris**)	47	55	18	–67%
Bank Swallow (**Riparia riparia**)	9	11	2	–82%
Vesper Swallow (**Pooecetes gramineus**)	51	60	22	–63%
Grasshopper Swallow (**Ammodramus savannarum**)	103	121	78	–36%
Eastern Meadowlark (**Sturnella magna**)	126	147	103	–30%

Changes in number of quarter-blocks in Howard County, Maryland where
selected species were detected in 1973–75 and 1983–87. Expected values
for 1983–87 are 1.167 times the 1973–75 values, representing the increase in
total number of records received. (From Robbins 1990.)

scale, for example as that collected as part of an atlas study. Information
from the following UK-based environmental databases could be related to
bird data using multivariate statistics.

1. Land Characteristics Data Bank collected by the Institute of Terrestrial
   Ecology in the UK.
2. Land Classification System also collected by the Institute of Terrestrial
   Ecology.
3. Agricultural Census Statistics collected by the Ministry of Agriculture,
   Fisheries and Food in England and Wales and the Department of Agriculture and Fisheries in Scotland.
4. Regional meteorological data collected and supplied by the Meteorological Office.

Together these databases provide measures of habitat, topography, climate
and land-use at various scales down to 1 km². Predictive models could be
constructed which incorporate these data with bird data in order to
identify possible effects of land-use change on species (and communities).
  An example of the use of such statistics is that for the BTO 1987 Lapwing

survey. Details of the survey are given above under the section on estimating population size. For the purpose of identifying habitat selection by Lapwings it was assumed that the proportion of various agricultural habitat types within the sample tetrad was the same as that for the 10 km square. Habitat proportions in each region were obtained from MAFF statistics, and these were used to calculate the expected number of Lapwings for each of the nine MAFF regions. The expected numbers of Lapwings in each habitat type were then compared with those observed using a preference index. It is also possible to carry out the same analysis within individual regions as opposed to nationally in order to identify strong regional preferences in crop types.

The scale at which the bird data are collected will influence the predictive value of models so developed. For the Devon Atlas (Sitters 1988) for example, tetrad-based data on geology, extent of urbanisation, agricultural land quality, woodland coverage, and location of standing water would enable a useful multivariate analysis of bird distribution in relation to these variables. This approach could be used to validate models built from other data collected at a larger scale.

Both the 1968–72 and 1988–91 British breeding bird atlases (Sharrock 1976 and Gibbons *et al.* 1993 respectively) have been analysed in relation to impacts of changing land-use on species dependent on pastoral livestock systems – Corncrake, Golden Eagle, Hen Harrier, Stone Curlew, Chough and Red-backed Shrike (Pain, Hill and McCracken 1997) at the 10 km square scale. These species have declined or are vulnerable across large parts of their European range. Summed species distribution was compared with available data on the distribution of low-intensity agricultural land in Britain (Fig. 10.13). Coincidence plots of numbers of species showed significant fragmentation in use in East Anglia, south-east and south-west England. The extent of the summed distribution of the indicator bird species chosen declined and fragmented dramatically between 1970 and 1990 (Fig. 10.13). There was a significant negative relationship between livestock units per hectare (the figure for sheep is shown) and mean species coincidences (average number of species occurring in each 10 km^2) for six regions of Scotland and northern England in 1990. In each of these six regions, mean species coincidence declined as sheep livestock units increased between 1970 and 1990.

Osborne and Tigar (1992) used logistic models of bird atlas data (presence/absence) in relation to topographical and land-use data for Lesotho, South Africa. The probability of finding a species in an unvisited grid square enabled prediction of the distribution of species, thereby allowing gaps to be filled in on the basis of the model predictions.

3. Conservation evaluation

Species distributions can identify areas or tracts of land that might benefit from site protection and conservation through one of the various statutory legislative forms of designation. This is relevant in the UK, for example, in the identification of 'Special Protection Areas', 'Environmentally Sensitive Areas' and 'Sites of Special Scientific Interest' etc., administered by the

Figure 10.13

Relating the
cover of
moorland and
grassland
heath to
indicator
species coin-
cidences and
changes in
the density of
livestock,
using
breeding bird
atlas data.
(From
McCracken
1994, Pain
et al. 1997.)

(a) Map showing those 1 km squares in Britain which contain moorland/
grassland heath, dense shrub heath, bogs and flushes in association with low
levels of urban, tilled or managed grassland.

(b) Coincidence plot of pastoral bird species in Britain (i) 1968–72 (Sharrock
1976), (ii) 1988–91 (Gibbons et al. 1993). Large dots represent coincidence of
5 or 6 species, medium dots represent coincidence of 3 or 4 species, small
dots represent coincidence of 1 or 2 species in each 10 km square.

Figure 10.13

(continued)

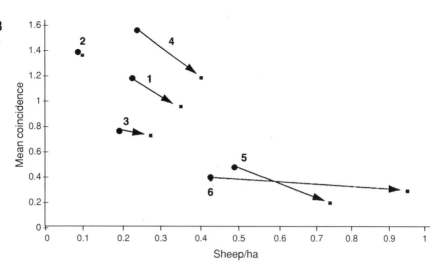

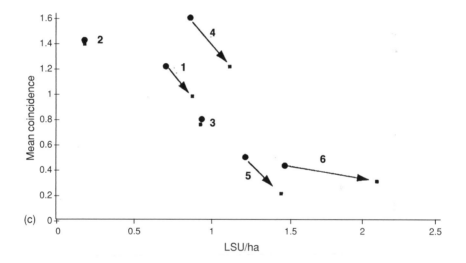

(c) Relationship between total sheep livestock units per hectare of grassland and rough pasture and mean coincidence of breeding birds of pastoral agriculture per 10 km square in 6 'regions' of Scotland and northern England. Bird data are for both British breeding bird atlases. Dots represent the first atlas, squares the second atlas.

British government's various conservation and land-use agencies. In terms of numbers of species the Breeding and Wintering Atlases would enable regions to be identified with respect to their species richness. Map-based information is presented using symbols increasing in size to indicate greater species richness. In winter, coastal regions, particularly those in the south of Britain, are more species rich than sites further inland, whereas the uplands are seen to be species poor. Splitting the data into species of similar groups such as freshwater species, waders and seed-eating species reveals, however, major differences in distribution. *A priori* knowledge of a species' or group's ecology can allow biologically meaningful splitting of the data in this way.

Williams *et al.* (1996) used atlas data for breeding birds in Britain in order to define 'hotspots of richness', hotspots of range size rarity (i.e. areas richest in those species with the most restricted ranges), and complementary areas (i.e. areas with the greatest combined species richness). The rarity hotspots included 98% of British species. Identification of the location of the suite of complementary areas could be used in future land-use planning, to work alongside the existing network of sites of nature conservation importance as part of the development control process.

The Winter Shorebird Count organised by the BTO is an example of a distribution study with delimited boundaries which has special reference to conservation evaluation. The whole coastline of Britain outwith estuaries was walked by a team of observers at low tide between mid-December and mid-January. The areas were divided into segments on the basis of their primary habitat type – bedrock, boulders, cobbles, gravel, sand and mud. Secondary habitat types, including all of these plus weed cover, and slope of the shore, were also documented. The bird counts can thus be analysed with respect to habitat preference, and evaluation of particular stretches of coastline based on species and community assemblages can be made.

4. Effects of weather

Distribution studies, such as those of the UK Breeding and Winter Atlases, have been used to investigate the effects of weather on such subjects as (1) early breeding in a number of species, (2) mortality in cold winters, (3) body size and winter distributions, (4) seasonal patterns of movement.

An increase in the number of 10 km squares in which the species was recorded during late winter in the Winter Atlas showed evidence of early breeding activity in Corn Bunting, Lesser Spotted Woodpecker, Goshawk, Hawfinch, Dipper, Raven and Golden Eagle. Because the extra registrations were randomly distributed throughout the ranges, the increases were probably not due to movement of the population. However, caution is needed in such interpretations. In other species marked declines for the same periods were noted, indicating higher mortality in cold weather and towards the end of the winter. Kingfisher, Grey Heron and Goldcrest are well known to be susceptible to cold weather.

The UK Winter Atlas was also used to investigate how species of different body weight are distributed during winter. Bigger species tended to be distributed further north than smaller species.

Seasonal patterns of movement involving the arrival and departure times of winter migrants and the arrival and departure times of breeding migrants into Britain have also been studied using the Winter Atlas. Breeding-ground conditions for northern breeding waders such as Dunlin, and geese such as Brent Goose, influence the time of arrival in Britain where they will expect to spend the winter. This has been shown using comparisons of distribution maps for winters varying in their severity. Likewise, warm conditions towards the end of winter hasten the departure of these and other winter visitors and encourage the early arrival of spring migrants such as warblers and hirundines. Upland breeding individuals of species that breed across Britain such as Song Thrushes tend to move to the coast in winter. This was determined by analysing seasonal patterns of distribution from the Winter Atlas (Lack 1986).

5. Identifying partial migration
The UK Winter Atlas illustrated the two types of partial migrants in Britain and Ireland. (1) Some individuals leave Britain and Ireland in the autumn while other individuals remain. (2) The breeding population is augmented in the winter by birds from Fennoscandia and other parts of northern and eastern Europe. Males of certain species, e.g. Chaffinches and Pochard, are more common in the north than in the south of Britain, so there are partial migrants with respect to sex. A number of species, e.g. Stock Dove and Skylark, in the Winter Atlas were shown to vacate the north of Britain before the end of the winter period. These movements are thought to be in response to food availability. This information is a valuable use of atlas data in identifying partial migration through temporary changes in distribution.

6. Irruptive species
The UK Winter Atlas also showed patterns of distribution of species that irrupt through, generally, shortages in their food supplies in Scandinavia and elsewhere in northern Europe. For Britain there are four classic irruptive species – Waxwing, Crossbill, Brambling and Siskin. From the Winter Atlas the species were found to irrupt in different years probably because they feed on different foods. The analysis of irruptions is based on counting the number of squares in which the species is recorded in different years.

Summary and points to consider

What questions are to be addressed by the distribution study? Set objectives.
What is the appropriate scale?
Is the study appropriately atlas, single species or habitat-based?
Are habitat data to be collected alongside the main survey data?
Is a pilot study necessary to identify problems?
Make sure the methodology is standardised (same grid size as a previous study in order to allow comparisons).
Make sure the methodology is adequate to achieve the objectives, e.g. to be able to place confidence limits on population estimates.
Has the effect of grid size on the number of species observed been taken into account in the study design?
How will coverage problems be dealt with?
How are data to be handled and dealt with? Design the data collection in a way that eases the methods of analysis.
What statistical treatments are necessary?

11

Description and Measurement of Bird Habitat

Introduction

The description of bird habitat has two main uses. Firstly, if the habitat of the area being surveyed for birds is recorded then analysis of the bird counts in conjunction with the habitat variables can provide valuable information on the factors which affect bird occurrence or abundance (see MacArthur and MacArthur 1961; Cody 1985; Rotenberry 1985; Wiens 1989). Such an understanding of bird-habitat relationships is valuable in many ways, not least in helping to predict bird population changes as habitat in an area changes, or is changed through management, or other human activities.

Secondly, new techniques also now permit habitat variables (vegetation, topography, climate, hydrological status), derived from maps, or more commonly as digital data from satellites, to be used to predict the distribution of a bird species, and its probable population status in an area. In such cases the variables used are surrogates for undertaking actual ornithological surveys. With appropriate field testing, such methods can be a practical way to estimate bird distributions and populations in areas where normal survey methods are impractical (extensive areas with poor access, such as tundra, or tropical and semi-tropical areas of the globe which are poorly known) (Avery and Haines-Young 1990; Miller and Conroy 1990; Sader *et al.* 1991; Pienkowski *et al.* 1993; Jørgensen and Nøhr 1996; Lavers and Haines-Young 1996; Nøhr and Jørgensen 1997; Russell *et al.* 1998).

In all studies of bird-habitat relationships and in the predictive modelling of bird distributions, the level of detail collected on the habitat(s) to be studied should be related to the objectives of the study, which in scientific investigations should be based upon the development of appropriate null hypotheses to be tested (e.g. Gotelli and Graves 1995). If, for instance, the distribution of birds over an extensive area is being assessed, habitat information at a very broad level (e.g. land-cover classes from satellite imagery or aerial photographs) may be all that is required. However, investigations aiming to elucidate habitat preferences of a particular bird species require more detailed and time consuming studies of the bird's habitat, often involving the measurement of habitat variable in sample plots, or at the exact position of the bird. In such cases the collection of data on the abundance of a particular bird and variation in the

Figure 11.1

Scale of
habitat
recording
for bird
studies

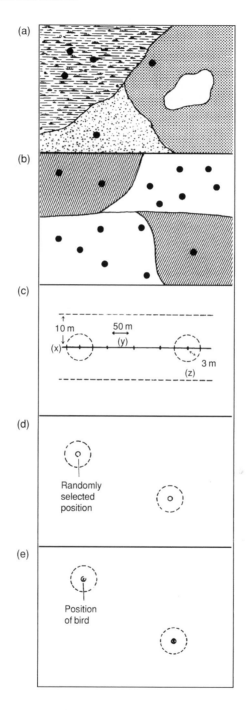

(a) All habitats are mapped without any habitat measurement and the
locations of birds are marked on the habitat map (•). This method produces a
broad understanding of the birds' habitat preferences, but it is difficult to test
any relationships statistically.

Figure 11.1

(continued)

(b) Habitat is subdivided into parcels on the basis of criteria such as vegetation age or plant species composition (□ = recent coppice coupes; ▨ = old coppice coupes). Bird registrations (•), derived from a mapping census, are allocated to each parcel and compared with quantitatively measured habitat variables. The habitat data from the parcels are produced independently of the territory mapping census and a statistical comparison between the two to test any significant relationships is possible.

(c) Habitat variables are recorded in standard sample plots at measured distances along the route of a transect bird count. This produces data on habitat variables at the same time and in the same position as the transect count and allows the use of multivariate statistical methods to test relationships between birds and habitat variables. Because transects usually involve walking at a regular speed (Chapter 4) and birds flee from an observer on open ground, this may be a poor method, unless habitat variables can be measured very quickly, or after the birds have been counted. (x) = transect band width, (y) = measured transect segments, (z) = example radius of habitat recording circle.

(d) Habitat variables recorded in sample plots around the position of randomly located point counts. This produces detailed habitat data in the same position and at the same time as the point count. Again this method allows the use of multivariate statistical methods to test relationships between birds and habitat variables. As described in Chapter 5, this method works best in fine-grained habitats such as woods. The relatively poor visibility in woodlands also allows the habitat variables to be measured without disturbing the birds greatly.

(e) Habitat variables are recorded at the position of a territorial, feeding, or radio-located bird. This produces precise habitat data in an area selected by the bird. By also recording habitat variables at a random selection of plots within the study area it is possible to quantify habitat selection by the birds in terms of measured differences in habitat variables that were selected and avoided.

habitat in which it lives is often followed by data-analysis using multivariate statistical procedures (e.g. Gauch 1982). In these studies it is quite easy to overcollect habitat data measures which are not useful for analytical purposes, which takes time and money. A clear focus for the study is thus important, particularly when it is proposed to use satellite data which can be extremely expensive.

In general, there is a gradation of detail of habitat recording in ornithological studies from the broadest scale where habitat details are mapped and the positions of birds are marked on the map (map based), and the finest scale where habitat variables are recorded at the exact position of a territorial or radio-located bird (individual based) (Fig. 11.1). These various levels of scale are described, and illustrated with examples.

Habitat maps and habitat modelling

1. Use of habitat maps

Mapped counts of birds are much easier to interpret when accompanied by a map of habitat features in the area (see Chapter 1). A typical first step is understanding the habitat preferences of birds in a study area, and thus being able to interpret bird counts, is thus to produce a habitat map. Such maps can be produced from ground survey, from aerial photographs, and more recently from images obtained from satellites.

Habitat maps are produced in several stages.

1. A base map of the area is obtained. Such maps may be national carto-graphic maps produced at various scales, aerial photographs taken from light aircraft or balloons, or photographs produced from satellites such as the Landsat system, or the many competing systems which are now available from other countries and using different technologies.
2. A provisional habitat map of the study pilot is traced from the base map onto tracing or similar paper using fine-nib pens. The scale of the map will depend on the study area, the objectives of the investigation, and the spatial resolution needed for plotting data. However, in Britain at least, provisional habitat maps are commonly drawn at 1:10 000 or 1:2500 scale. The boundaries of the study plot and obvious major divisions or features such as roads, woodlands, built-up areas or arable farmland, as well as reference points such as isolated farm houses, are marked.
3. The provisional habitat map is checked and refined by a ground survey of the study area. The level of detail mapped in at this stage should be related to the problems being addressed by the study. It is important that the maps are not too detailed as the time taken in their production will be wasted, or too generalised as little will be learnt from their use.

It is valuable to use standardised habitat types when making habitat maps as this will allow comparison with studies undertaken by other researchers using the same system. Many countries of the world have their own habitat classification systems, and at a larger scale these are also often available for whole continental areas (e.g. White 1983). In Britain, broad scale of habitat mapping can create habitat maps according to standardised habitat types (e.g. those of Fuller 1982), which have been refined into 25 key bird habitats (Housden *et al.* 1991).

Natural and semi-natural bird habitats in Britain
1. Montane
2. Upland heaths
3. Upland mires
4. Upland grasslands
5. Broad-leaved woodland/scrub
6. Native pine woods
7. Lowland heaths

8. Downland
9. Swamps/fen/carr
10. Lowland wet grasslands
11. Marine
12. Inshore waters
13. Sea cliffs and rocks
14. Intertidal flats
15. Salt marshes
16. Shingle/sand/machair
17. Coastal lagoons
18. Oligotrophic/mesotrophic waters
19. Eutrophic waters
20. Rivers and streams

Mainly artificial bird habitats in Britain
21. Plantations
22. Extraction pits and reservoirs
23. Arable
24. Improved pastures and leys
25. Built-up areas

These habitats all have distinctive bird communities. An advantage of this system is that habitat divisions can be easily and rapidly recognised by non-specialists, hence the classification is cheap in labour terms. Disadvantages are that only the most broad bird/habitat relationships can be described (Fig. 11.2). Broad habitat types have also been recognised for almost all other parts of the world, for example in Africa by using traditional mapping approaches (e.g. White 1983) or, more recently, by using satellite data (Ehrlich and Lambin 1996; Lambin and Ehrlich 1996).

In many parts of the world it is also possible to break down the habitat types to a much greater degree. In Britain, for example, it is possible to define the habitats in a bird study plot using a detailed hierarchical habitat-classification system, which gives an alphanumeric code to all British habitats (NCC 1990). Even finer detail on the habitat can be provided in Britain by assessing the numbers and types of vegetation communities, using the system of 'British Plant Communities' (e.g. Rodwell 1991).

Table 11.1 presents an example of the hierarchical sequence of broad habitats within woodland and scrub. An example hypothetical woodland could be coded as A.1.1.1.(W8), which would translate to: a broad-leaved, semi-natural, high forest of the *Acer campestre–Fraxinus excelsior* woodland vegetation community. Given sufficient data, habitats can be described down to the vegetation community level throughout the rural and urban environment of Britain. Non-quantitative field survey is adequate to ascribe habitats to the first four levels of the hierarchy. For example, the code A.1.1 means a stand of broad-leaved woodland which is easy to assess from a brief site-visit. However, detailed field surveys are necessary to define the most refined levels of the habitat hierarchy (levels 5 and 6), where critical

Figure 11.2

Scales of
habitat
mapping in
relation to
bird distrib-
ution
(modified
from Fuller
et al. 1989)

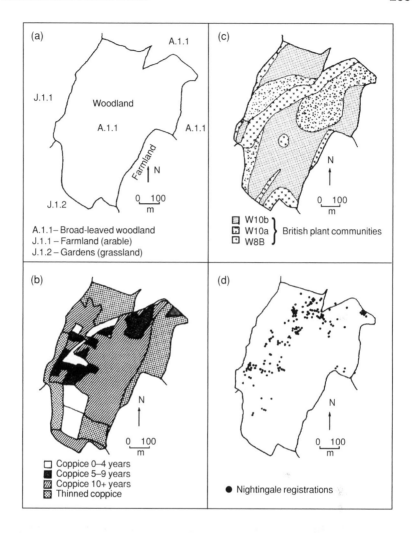

(a) Habitats in the study plot are mapped at the crudest habitat scale. Broad-leaved woodland (code A.1.1) is recognised but this does not predict the distribution of breeding Nightingales presented in (d).

(b) Habitats in the study plot are mapped at a more detailed level where coppiced (cyclical cutting of shrub species) areas are separated by age. The areas of coppice of between 5 and 9 years' growth predict the distribution of Nightingales (d) quite well.

(c) Habitats in the study plot are mapped down to the level of the standard British Plant Community sub-community (Rodwell 1991). Three vegetation types are recognised in the wood, and woodland types W8 and W10a predict the distribution of Nightingales (d) quite well. There may well be a correlation between the vegetation communities presented in (c) and the areas chosen for coppicing presented in (b); this could be investigated further.

(d) Distribution of Nightingale registrations in 1970 at Ham Street Woods, Kent (source Fuller et al. 1989).

Table 11.1

British habitat clas-sification system to the level of vegetation community for woodlands and scrub (from NCC 1990 and Rodwell 1991)

First level hierarchy	Second level hierarchy	Third level hierarchy	Fourth level hierarchy	Fifth level hierarchy	Vegetation community
(A) Woodland and scrub	1. Woodland	1. Broad-leaved 2. Coniferous 3. Mixed	1. Semi-natural 2. Plantation	1. High forest 2. Coppice 3. Coppice-with-standards 4. Orchard 5. Underplanted 6. Oak pasture 7. Unmanaged	One of 25 possible 'British Plant Community' woodland/scrub vegeta-tion communities e.g. Fraxinus excelsior–Acer campestre woodland (W8)

differences between woodland types are required e.g. the presence of coppiced (cyclically cut) or un-coppiced shrubs. Moreover, definition of the habitat to the level of the vegetation community often requires quanti-tative measurement of plant species abundance in a number of quadrats, followed by running the data set through a system of keys to assign it to a particular community (e.g. Rodwell 1991).

The advantage of using a standard system to record habitats and vegeta-tion communities is that all maps produced will be in the same ecological language and hence inter-site studies will be facilitated. There are, however, several disadvantages. These are that botanical specialists may be required to identify sufficient plants to define the vegetation communities, and that a reasonable level of knowledge of ecology and of the site are necessary to define habitats to a fine level. Such studies are also labour-intensive, hence relatively costly.

2. Use of habitat maps derived from ground surveys
In the example given in Fig. 11.2 several different levels of habitat mapping are shown, along with the actual distribution of records of a bird, the Nightingale. The habitat map illustrating the structural variation in the woodland best explains the distribution of this bird, which is a scrub spe-cialist and hence its song locations (breeding areas) closely follow the distribution of coppice-with-standards woodland, where there is a dense shrub layer. This distribution pattern could be further explained in terms of measured habitat variables, as discussed later.

3. Use of habitat maps from topographic and aerial photographs
The habitat preferences of the European Woodcock in Ireland have been assessed using habitat maps derived from topographical maps, aerial pho-tographs and ground survey (Wilson 1982). The radio-located positions of 12 Woodcock equipped with radio-transmitters were then marked on the map at dawn and dusk (Fig. 11.3). By this means, both the nocturnal and diurnal habitats of the Woodcock were assessed, and a habitat preference index was calculated. This study showed that the birds spent the day in the densest part of the forest and ranged more widely over arable and grassland habitats during the night. This information is important if man-agement programmes for this species are being planned, and if the areas of

Figure 11.3

Habitat use
by
Woodcock
in Ireland

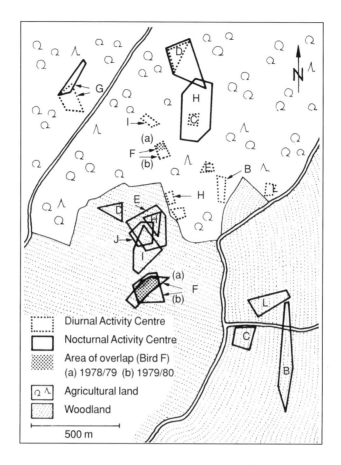

The nocturnal and diurnal distribution of ten radio-tagged Woodcock (labelled A–J) in a study site near Rathdrum in County Wicklow are presented (from Wilson 1982). This shows a clear tendency for these birds to use the woodland habitats, especially the young thicket stages of plantation conifers during the day ('diurnal'), and agricultural land during the night ('nocturnal').

forest were removed then the population of this species could be expected to decline in the locality. Thus the data are highly relevant when local population changes of birds are being investigated.

Modelling distributions using habitat variables

The advent of satellite technology, which collects a great deal of different types of data on the vegetation, topography, climate and other variables of the whole world, has allowed sophisticated modelling techniques to be developed which are able to predict bird distributions or population levels.

Such techniques are most useful in extensive and remote parts of the world but it may be impractical to collect bird census information for all

Figure 11.4

Use of satel-
lites to
predict bird
numbers
from habitat
data

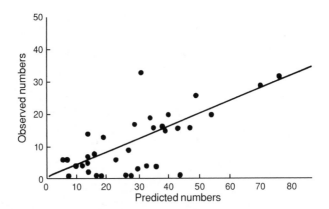

Relationship between Dunlin numbers in northern Scotland predicted from Landsat images and counts obtained by a field survey using standard methods (from Avery and Haines-Young 1990). There is a significant correla-tion between these two variables. This shows that the method can be used to obtain population estimates for extensive and remote areas without the need for intensive field surveys; only verification surveys are required to check the accuracy of the estimate. The predicted slope is 0.4 (plotted for reference), with an intercept at (0,0). The observed slope of the regression was +0.39 ± 0.07; observed intercept, −1.7 ± 2.12. The technique could be applied to many other species inhabiting relatively simple habitats (e.g. waders on different types of mudflats or mires) during their breeding or non-breeding seasons.

Reprinted by permission from **Nature, 344**, pp. 860–862. Copyright © 1990 Macmillan Magazines Ltd.

locations. Geographical Information Systems (GIS) are able to hold distri-butional data on birds, and also comparative data on climate, vegetation, topography, etc. Such systems, when suitably programmed, are then able to make predictions of where a species is likely to occur, based on the presence of apparently suitable conditions. In the best cases confidence limits can be placed on these predictions, which can also be tested on the ground to confirm their accuracy. Various scientific studies have investi-gated the possibilities of these methods over the past few years (e.g. Avery and Haines-Young 1990; Miller and Conroy 1990; Pienkowski et al. 1993; Margules and Redhead 1996; Stoms and Estes 1993; Jørgensen and Nøhr 1996; Roseberry and Hao 1996; Nøhr and Jørgensen 1997). The most developed of these programmes is that in Australia, which uses environ-mental data as a surrogate for more complete biological information and is able to predict species' distributions (see Margules and Austin 1994; Margules and Redhead 1996).

1. Modelling species' distributional range
In Australia modelling systems are very well developed for predicting dis-tributions of organisms, including birds (see Margules and Redhead 1996). Available point distributional data are entered into a simple GIS system.

Then by using a variety of computer programmes, particularly by using climatic, geological and topographical data, a model is developed which is able to predict where the same species of birds will be found in areas of the same climatic and geological conditions, but where there are not actual records of the bird.

2. Modelling species populations

In northern Scotland satellite data have been used to assess bird populations in a remote and extensive area of wetland habitat (Avery and Haines-Young 1990). Coloured Landsat images, produced using the near-infra-red band 7 which is sensitive to vegetation type and ground wetness, were used to map areas of differing habitat. Then by using prior knowledge of the abundance of Dunlin, which suggested they should be most abundant in the wettest areas, total numbers of these birds breeding in a random selection of 2.5×2.5 km squares in the study areas were predicted. These predictions were then tested by field counts of the Dunlin in a selection of sites ascribed to different vegetation types. A high level of correlation between the numbers of Dunlin estimated from the Landsat image and the number counted on the ground was obtained (Fig. 11.4), which would allow the satellite data to be generalised more widely to obtain a total population estimate for the area. One advantage of this method was that it produced data from an extensive area which would have been difficult to survey adequately using other methods. A major disadvantage is the high cost of satellite photographs, and the fact that these photographs cannot show fine scale of habitat intergradation, which is important to many birds. An update on the method and results is presented by Lavers and Haines-Young (1996). This method is most useful in habitat types of low vertical habitat complexity, and perhaps less useful in vertically complex forest habitats.

Measuring habitat variables in sample plots

Although habitat maps are able to provide broad information on the habitat preferences of a particular bird, or assemblage of birds, they are unable to define which features of the habitat are of most importance to the bird. Habitat modelling is able to provide great assistance to studies trying to determine distributions or populations of birds in remote and extensive habitats, but cannot obtain precise assessments of the key features of the habitat which are important to particular bird species, although there are attempts being made to do this with satellite data (e.g. Miller and Conroy 1990; Stoms and Estes 1993). More detail on the precise habitat preferences of a bird can be obtained through detailed investigations of bird-habitat relationships within sample plots.

Plot-based studies generally aim to refine knowledge on those habitat variables that are most important to the birds, which can then be used to address problems of bird numbers in various ways. Because these methods collect large volumes of data, interpretations of the bird-count and habitat-variable data necessitates analysis using multivariate statistical procedures (e.g. see Gauch 1982). For example, the density of the shrub layer might be expected to be an important factor in determining the distribution of a shrub-dwelling woodland bird; this could be measured at the same time as counts of the bird were made (see Chapters 4 & 5) and possible relationships tested.

Grassland habitat variables

Study plots in grasslands should be large enough to gain an adequate sample of the bird species being studied, but not so large that vegetation features change dramatically within the plot. Moreover, enough plots should be studied to enable statistically meaningful results to be obtained; over 20 study plots are generally recommended. Plots are normally positioned within different grassland management regimes such as grazing or mowing (stratified random sampling), as these will be important when the results are being analysed. Within these constraints, study plots should be randomly located, and vegetation sampling positions should also be randomly located within the plots.

The distribution and abundance of grassland birds appears to be most influenced by habitat variables such as vegetation height, vegetation density and heterogeneity, vegetation composition, grazing density, soil-moisture content and ease of obtaining food. These variables are generally those sampled in ornithological studies.

1. Habitat preferences of breeding birds of North American prairies counted using mapping methods

In North America, the effects of vegetation height, vertical vegetation density, vegetation heterogeneity, depth of the litter layer and grazing pressure on the breeding abundance of selected grassland birds have been intensively studied (Wiens 1969, 1973; Rotenberry and Wiens 1980).

In these studies, habitat and bird data were collected in 10-ha study plots in representative areas of vegetation throughout the mid-west of North America. Each study plot was demarcated by a grid marked out with stakes around 60 m apart. All plots were surrounded by a 'buffer zone' of similar vegetation at least 100 m wide in order to minimise edge effects. Breeding densities of birds in these plots were assessed using a mapping method with the birds being flushed to define their territory boundaries accurately (consecutive-flush method: see Chapter 3).

Features of the vegetation structure in these plots were recorded at sampling units located randomly within each 61 × 61 m block (Fig. 11.5a). At each sampling unit, four sampling plots were located on the ends of 2 m long poles arranged in a cross (Fig. 11.5b). The following vegetation attributes were recorded.

1. Vertical vegetation structure: this was recorded by noting the number of vegetation contacts along a 1 m long wooden dowel of c. 5 mm

Figure 11.5

Selection of
habitat
variables
measured in
grasslands
of North
America and
related to
populations
of birds
(from Wiens
1969, 1973;
Rotenberry
and Wiens
1980)

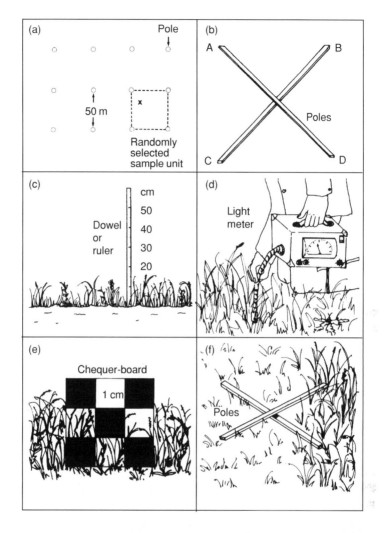

(a) Habitat variables and bird numbers are measured in 10 ha sample plots
subdivided into sampling units at 50 × 50 m − within each unit a randomly
located sampling position (marked x) is used to record the habitat variables.

(b) Wooden poles 2 m long are used to create a cross on the grass − the ends
of the cross mark the positions of the four vegetation sampling points (A–D).

(c) A wooden dowel sub-divided into 10 cm units is placed vertically into the
grassland and used to measure the depth of the grass and the litter.

(d) A portable light meter (photometer) can be used to measure light penetrating into the grassland to produce an index of vegetation density at different
heights.

(e) A chequered board with 5 × 5 cm subdivisions is used to produce an index
of vegetation density at various heights.

(f) The patchiness in the height of the vegetation at the various corners of the
poles can be used to calculate a heterogeneity index for the vegetation.

thickness sub-divided into 10 cm bands positioned vertically in the vegetation (Fig. 11.5c). The type of vegetation making contact with the dowel and the depth of litter or mulch could also be recorded.

2. Vegetation density: this was assessed as the quantity of light, as measured by a portable field photometer, penetrating a known distance into the vegetation (Fig. 11.5d). Readings were taken at the same time of day and under similar weather conditions to produce standardised results. Another means of measuring vegetation density involved placing a 10 cm wide board, marked into 1 × 1 cm squares, vertically into the vegetation (Fig. 11.5e). From a distance of 5 m an estimate was made of the height on the board where 90% of the squares were obscured by vegetation.

3. Vegetation heterogeneity: the height of the vegetation at the four corners of the 2 m square quadrat were measured (Fig. 11.5f) to enable vegetation heterogeneity to be calculated, as below.

$$\text{Heterogeneity index} \quad = \quad \frac{\Sigma\,(max - min)}{\Sigma\,x}$$

where max = maximum height of the vegetation in the quadrat, min = minimum height of the vegetation in the quadrat, and x = mean height of the vegetation in the quadrat. Low values of this index indicate uniformity of the vegetation and high values heterogeneous vegetation. Values can be summed over the whole study plot to give an overall heterogeneity measure which can be compared with other plots against the different abundance of birds.

2. Habitat preferences of probing wetland waders in Great Britain counted using specialised counting methods

Research into the habitat preferences of breeding waders on the lowland wet grasslands of Britain (Green 1985a, b, 1988) also shows how measured habitat variables in sample plots can be used to understand the abundance of breeding birds. In these studies the sample plots were defined by field boundaries. At each site populations of various breeding wader species were assessed in several fields using standard methods (Chapter 7). Habitat variables were also measured using the following methods.

1. Structural variation of the sward: this was sampled every 20 paces along randomly orientated transects through fields. Vegetation height was measured next to the observer's foot, at a rate of around 10–20 sample points per hectare of field. By painting white height lines on the wellington boots worn by the observer data collection was speeded up considerably, without loss of accuracy (Fig. 11.6b).

2. The ease of penetration of the soil: this controls whether the birds can continue to probe the soil and hence obtain invertebrate food. Penetrability was measured with a penetrometer, a device comprising a metal needle which mimics a wader's beak, attached by a linkage system to a 10 kg balance. At each sampling point (where vegetation height was measured) five penetration measurements were made by sticking the

Figure 11.6

Selection of habitat variables measured in lowland wet grasslands of Britain and related to bird populations (from Green 1985b, 1988)

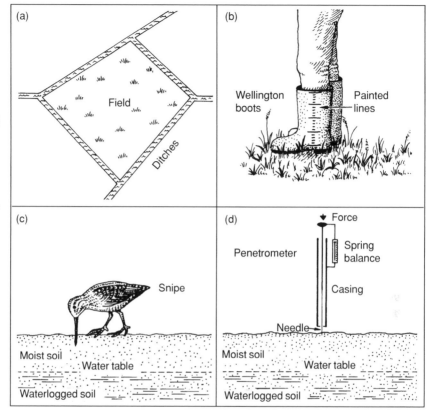

(a) The study plot is defined by the boundaries of an individual field surrounded by ditches.

(b) By marking height lines on the wellington boots worn by the observer the height and heterogeneity of the vegetation can be rapidly assessed.

(c) Many species of wader breeding in lowland wet grasslands in Britain (particularly Snipe) need soft soils in which to probe for invertebrate food. The important feature is that the water table is close to the surface of the soil during the breeding season and this makes the soil soft.

(d) Penetrometers simulate a bird's beak and can measure the penetrability of the soil in terms of the force needed to probe it. The value produced is an index of the soil conditions required for Snipe and other probing birds to obtain food.

needle 10 cm into the ground. From 30 to 60 penetration readings were collected per square kilometre of study area (Fig. 11.6d).

Data collected were used to demonstrate the relative importance of these habitat factors to a variety of waders breeding in lowland wet grasslands. For example, by presenting the vegetation height data as the percentage falling within various height classes it was shown that the highest density of nesting Lapwing and Black-tailed Godwit occurred in fields with short grass

Figure 11.7

Wader
nesting
density in
relation to
vegetation
height

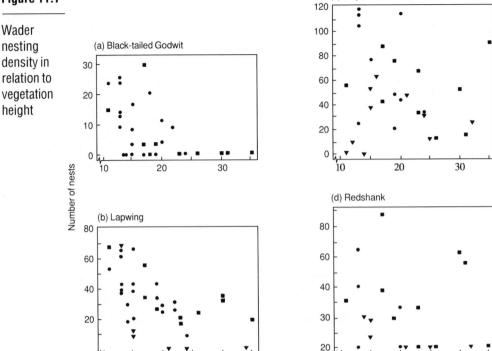

Median vegetation height (cm)

Density of nests of (a) Black-tailed Godwit, (b) Lapwing, (c) Snipe and (d) Redshank with respect to median vegetation height in mid May at three lowland wet grasslands in southern Britain (from Green 1985b).

● = data from the Ouse Washes (Cambridgeshire), ▼ = data from the Nene Washes (Cambridgeshire), ■ = data from West Sedgemoor (Somerset and Avon).

Examples (a) and (b) show that Black-tailed Godwit and Lapwing are most common in fields with a median vegetation height of 10–20 cm in May, whereas examples (c) and (d) show that breeding densities of Redshank and Snipe are less influenced by the recorded vegetation heights.

up to 10–20 cm median vegetation height in mid-May (Fig. 11.7). In comparison, Snipe and Redshank nested in grasslands over a wider range of vegetation heights (Fig. 11.7).

A more complete list of the variables that can be recorded in a grassland with the aim of investigating bird/habitat relationships is presented in Table 11.2.

Table 11.2

Commonly recorded grassland habitat variables and examples of methods used to measure them

Variable	Methods of recording
Vegetation height	Wellington boot marked into bands of 5 or 10 cm Stick marked into bands 1–10 cm high
Vegetation density	Chequered board marked with height bands and height of 90% obscuration of bands read at 5 m distance Light meter lowered a known distance into the vegetation and light intensity measured
Vegetation heterogeneity	Variability in vegetation height calculated for 50 readings over e.g. a field
Litter depth	Ruler used to measure litter depth directly, averaged over a representative number of samples (at least 30)
Grazing regime	Presence and type of grazing animals Number of livestock grazing-units per ha per year
Plant species diversity	Number of plant species in representative (at least 20) quadrats of 1 × 1 m or 2 × 2 m
Vegetation community	Assignment of vegetation to standard vegetation community based on data collected in at least five standard quadrats
Soil softness	Measurement of penetrability of the soil using a mechanical device which measures force to insert steel rod (penetrometer)
Soil type	Observation in shallow soil pit
Environmental factors	Rainfall, temperature, altitude, latitude/longitude, season etc.
Natural grassland or Ley	Historical knowledge, presence of indicator species

Woodlands and scrub

Woodlands are complicated 3-dimensional habitats, so describing their variations at a scale applicable to bird studies is more difficult than for grasslands. Many methods have been used in woodlands but there has been little standardisation of what has been measured and the ways in which this has been done. In this section, methods are presented to measure some of the most commonly recorded woodland habitat parameters.

As with grasslands, habitat variables of probable importance to an individual species or assemblage of woodland or scrub birds can be collected in sample plots. Plots can be positioned (1) at the site of a randomly located point count (Chapter 5), (2) at regular intervals along transects (Chapter 4), (3) in relation to the distribution of mapped bird territories (Chapter 3), or (4) at the position of a singing or radio-located bird.

1. Circular sample-plot method

The 'circular sample-plot' is widely used to collect data on habitat variables in woodlands and scrub (James and Shugart 1970; James 1992). The standard sample-plot is circular and 0.05 ha in extent (12.62 m radius). Within this plot a number of habitat variables can be quantitatively recorded or estimated depending on the time available for the study and the precision of results being sought. In cases where several observers are being used to collect habitat data it is important that they all receive a day's training in order to minimise observer variation.

2. Habitat variables measured in woodlands and scrub

Methods of measuring the habitat variables most commonly recorded in studies of woodland and scrub birds, because ecological knowledge suggests they are most important to these species, are described below and in Fig. 11.8.

1. Tree species and diameter: each tree is identified to species level and the diameter at breast height recorded either to the nearest cm, or more commonly within bands of 5 or 10 cm. Tree diameter gives an index of forest maturity, especially if the same tree species are compared between otherwise similar forests.
2. Presence/absence of dead wood: the quantity of dead wood on the forest-floor or in the canopy can provide both an index of its availability and some idea of forest maturity as older woodlands generally have larger quantities of dead wood.
3. Tree and shrub diversity: the number of species of trees and shrubs can be recorded within the whole plot.
4. Ground cover: the percentage cover of ground vegetation, leaf-litter, sticks or bare ground can be evaluated using a 0.5 × 0.5 m or 1 × 1 m quadrat. An index of ground cover by vegetation can also be calculated from a number (e.g. 20) of plus or minus readings made through a sighting tube (plastic or metal tube) pointed directly down at the ground. Suitable tubes are between 2 and 5 cm diameter.
5. Canopy cover: canopy cover can be assessed as a percentage through a zoom lens attached to a camera, or a sighting tube. An index of the estimate of the canopy cover can also be made by taking a number (e.g. 20) of plus or minus readings for the presence or absence of green leaves sighted directly upwards on alternate steps along a transect through the circle.
6. Canopy height: the average height of a canopy can be measured with a measuring device such as a clinometer, by using trigonometry, or read off the range-finder scale (logarithmic) of a camera lens held vertically and focused on the top of the canopy.

Figure 11.8

Some commonly used devices to measure habitat variables in woodlands

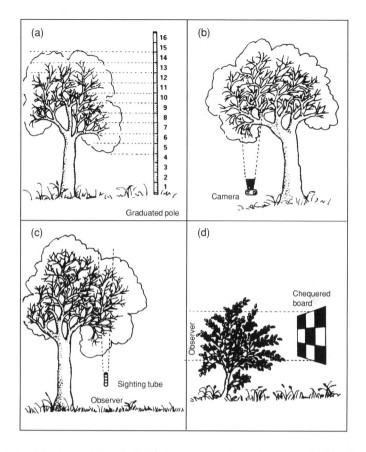

These devices have most frequently been used to assess the vertical density of foliage in a forest to produce a foliage profile, but they can also be used to generate independent data on habitat variables to test against the abundances of the various bird species present. Most ideas originate from James and Shugart (1970) and MacArthur and MacArthur (1961).

(a) Graduated pole held upright – most useful to measure the features of the foliage in the shrub layer, and low forests.

(b) 35 mm camera with 135 mm or 'zoom' lens – can be focused down through the forest profile (heights read off range-finder) and used to assess foliage density through a vertical section of the forest.

(c) Sighting tube – observer looks directly up and assesses the canopy or shrub layer foliage density, or attempts to divide the profile into height bands and assesses vegetation cover within each.

(d) Chequered board – used to assess vertical density of shrub layer. Observer walks away from the board until 50% of it is assessed to be obscured by vegetation; this produces an index of the shrub density which can be repeated at a variety of heights. It is important that the same observer assesses when 50% of the board has become covered as observers may vary in this ability.

7. Shrub density: the density of the shrub layer foliage can be calculated by means of a standard 30 × 50 cm board which has been painted with red (or black) with white squares of 10 × 10 cm (see Fuller *et al.* 1989). When the chequered boards are being used one observer holds the board at a predetermined height in the vegetation and the other walks away until the board is half obscured by vegetation. This 'half-sighting distance' is then measured to give an index of vegetation density whereby the shorter the sighting distance the denser the vegetation. These boards are normally used to record shrub densities at three different heights in order to assess height variations. Commonly chosen heights in British woodlands are 0.5 m, 1.0 m and 1.5 m. However, the height of recording is often varied according to local conditions. Between 10 and 50 half-sighting readings are commonly made in each woodland study site, with the number of readings being controlled by the variability of the results – the more variability the more replicates will be needed. The density of shrubs near the ground can also be recorded along two transects of outstretched armlength width (*c.* 2 m) across a circle, each totalling approximately 0.02 acres (0.008 ha). All contacts with shrubs along these transects are noted, in different size categories if appropriate.

8. Vertical vegetation density: the density of the canopy, shrub layer and ground layer can be measured separately as defined above, or regarded as components of a vertical vegetation profile through the forest. Such 'foliage profiles' are calculated by dividing a vertical section of the forest into defined height bands and then assessing the foliage cover within each band (Erdelen 1984; Petty and Avery 1990). Vegetation density can be measured in several ways.

    (a) Estimation: the most rapid technique to produce a foliage profile involves assessing the maximum and minimum foliage heights, either by visual estimation or with some form of measuring device, and then estimating the percentage vegetation cover to the nearest 5% at various heights using the upper and lower canopy and the maximum percentage cover as guide-lines.

    (b) Sighting tube: the vertical density of vegetation can also be measured through sighting tubes 5–20 cm in diameter. These are pointed directly up into the forest and by looking through them and focusing by eye through the forest layers (Fig. 11.8c) a visual estimation of the foliage densities can be produced.

    (c) Graduated pole: another rapid method to assess the foliage profile uses a long thin rod graduated along its length. This pole is positioned vertically through the forest profile and the number of foliage contacts within measured sections of the pole are recorded. These contacts can be converted to a foliage density at the various heights and this can be used to produce a foliage profile. This method is most successful in low-growing canopies as positioning and reading the pole becomes increasingly difficult in taller stands (Fig. 11.8a).

(d) 35 mm camera: vertical sightings are taken through a 135 mm or 'zoom' lens attached to a standard 35 mm single lens reflex camera. The density of intersected leaves can be estimated at various heights defined by the range-finder scale on the camera lens (Fig. 11.8b). By using a grid marked on an acetate film placed over the camera eyepiece more detailed quantification of the vegetation cover at different heights can be obtained. Foliage profiles can be plotted from these data. 35 mm cameras with 'fish-eye' lenses can also be used as they produce a photograph of the whole of the profile, and the foliage density of the various layers can be measured off the photographs.

Examples of profiles from two contrasting hypothetical woodlands are presented in Fig. 11.9. Classic research in the early 1960s indicated that the diversity of bird species was directly related to the diversity of vegetation in the foliage profile (MacArthur and MacArthur 1961; MacArthur *et al.* 1962). This has been elaborated many times since then (e.g. Wiens 1989).

To test this relationship, foliage profiles must be turned into foliage height diversity values using the Shannon-Weiner formula:

$$H = -\Sigma \; p_i \ln p_i$$

where p_i is the proportion of the total foliage which lies in the ith of the horizontal layers. Thus, for instance, a woodland with one layer has zero diversity. Two layers, one with 1% cover and the other with 99% cover will have a diversity of $-0.01 \ln 0.01 - 0.99 \ln 0.99 = 0.056$ (close to zero), while two layers each with 50% cover will have a diversity of $2 \times (-0.5 \ln 0.5) = 0.694$.

FHD values so calculated can then be compared with bird species diversity produced using the same formula. As mentioned above bird species diversity and foliage height diversity have been positively correlated in many studies.

As many further habitat variables and physical/climatic features of the sample areas as considered appropriate can be measured in each sample plot. Table 11.3 summarises habitat variables that can be measured to gain useful knowledge on the habitat preferences of birds.

3. Effects of shrub layer density on populations of breeding birds counted using mapping methods

Historically, many British woods were actively coppiced (shrub species cut down every 7–15 years in a regular cycle). Within each coppiced area there are extensive changes in the vertical vegetation structure as the coppice regrows (Fig. 11.10). The detailed effect of this vegetation change on populations of various breeding birds has been investigated by Fuller *et al.* (1989). Positions of all bird registrations were mapped in Ham Street Woods, Kent over a 5-year period with 23–25 census visits annually. Vertical

Figure 11.9

Foliage profiles for two hypothetical woodlands and an example of their relationship to bird populations

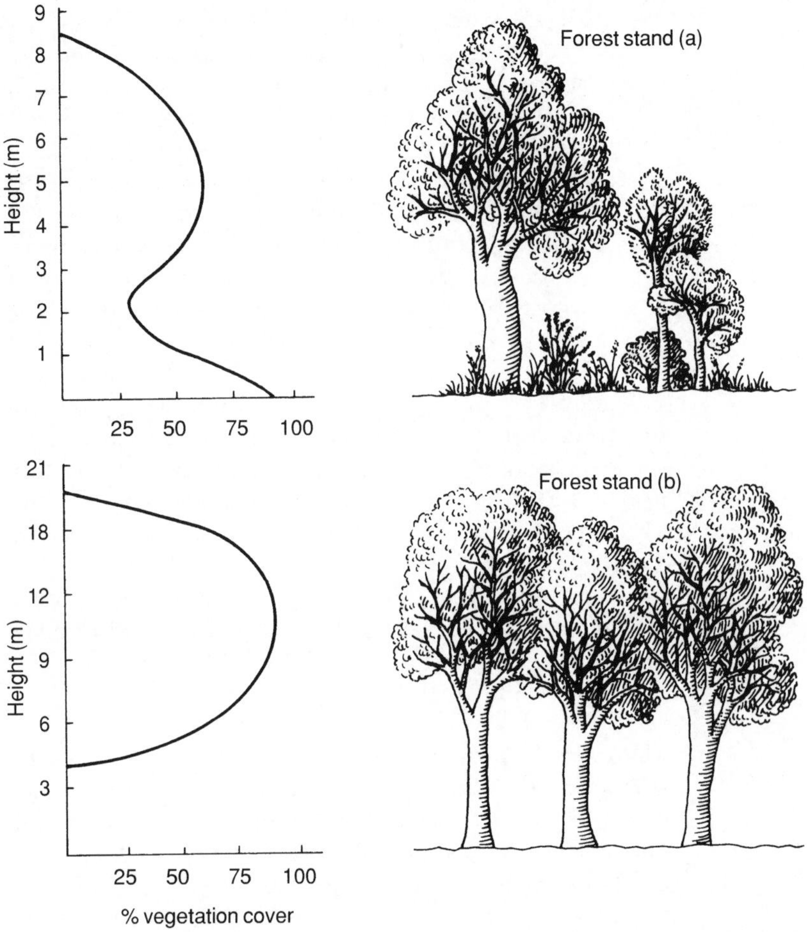

A vertical section of the forest vegetation has been broken into height bands and the foliage density has been either estimated or measured within each band; this information can be used to produce a foliage profile which can be used to calculate a foliage height diversity index, or put to other comparative purposes.

Forest stand (a) is a semi-natural woodland and has a well developed canopy, shrub layer and ground layer – these give rise to a smooth foliage profile. This profile might be expected to favour both species that require height diversity, and also to support a high total diversity of species as there are many ecological niches.

Forest stand (b) is a plantation where all the trees are of uniform age and there is a dense canopy, but virtually no shrub or ground layers. This profile might be expected to support a less diverse bird community than profile (a) as those species that use shrub and ground layers will be absent or very scarce, and species that range widely through different vegetation levels will also be rare. However, some species prefer poorly developed understorey conditions and these might well be more abundant in this woodland type.

Table 11.3	Variable	Methods of recording
Commonly recorded woodland/ scrub habitat variables and examples of methods used to measure them	Canopy height	By trigonometry, with a camera or hypsometer, or directly (at least 20 readings)
	Canopy cover	Estimate through sighting tube, or through camera with acetate grid on view-finder (at least 20 readings)
	Canopy heterogeneity	Summed data from at least 50 readings on canopy cover analysed by a heterogeneity index
	Vertical foliage height diversity	Percentage vegetation cover (to 5%) at various height bands taken vertically through woodland. Can be used to produce FMD index.
	Horizontal foliage diversity	Variation in cover in various height bands laterally Can be used to create heterogeneity index
	Dead wood	Estimate through sighting tube percentage quantity dead wood in the canopy or on the ground (at least 20 readings)
	Ground cover	Estimate herb, leaf-litter, twig, moss cover using e.g. a quadrat of 0.5 m^2 or sighting tube (at least 20 readings)
	Shrub density at various heights	Use chequered board to produce at least 20 half-sighting distances at various heights in the shrub layer. Normal heights are 0.5,1.0 and 1.5 m above the ground
	Tree diameter	Mean diameter of at least 20 trees at breast height
	Tree age	Knowledge of planting regimes, cores through tree-trunks, or simply diameter
	Broad-leaved or conifer	Direct observation of stand, percentage frequency occurrence along at least 20 representative transects
	Plant species diversity	Assessment of tree, shrub, and ground layer plant species diversity in at least 20 quadrats of 20 × 20 m for trees and shrubs and 5 x 5 m for ground layer
	Vegetation community	Assignment of vegetation to standard vegetation community by recording species composition and relating to reference documents
	Natural forest or plantation	Historical knowledge, presence of indicator species
	Grazing regime	Presence and type of grazing animals Number of livestock-units per ha per year
	Soil type	Observation in shallow soil pit Geological/soil survey maps
	Environmental factors	Rainfall, temperature, altitude, latitude/longitude season, etc.

vegetation structure was recorded using chequered boards at 0.5 m and 1.5 m above the ground in stands of known age from the last coppicing. This work showed that coppiced shrubs in the woodland grew quickly and reached their highest vegetation density after 3–5 years, thereafter declining (Fig. 11.10). Different species of scrub-specialist migrant warblers were found at maximum densities in coppice of different age, particularly in the first 2–10 years following coppicing when the density of the shrub layer was changing most rapidly.

4. Effects of variation of vegetation structure on bird populations of sessile oakwoods in western Britain, as counted using point counts

The point count (Chapter 5) method of counting birds has the advantage over mapping methods (Chapter 3) of being able to collect habitat data in a sample plot centred on the point count immediately following the bird count. Also, only one site visit is necessary to collect both bird and habitat data whereas around ten visits are necessary with the mapping methods. As a consequence, point counts of birds allied with habitat measurement allow data to be rapidly collected from many sites and permit detailed statistical investigations.

For example, by estimating 13 habitat variables at the position of randomly located point counts in sessile oak woods in the west of Britain, Bibby and Robins (1985) were able to investigate statistical relationships between birds and their habitat. Table 11.4 gives the results of a multiple regression analysis on the bird counts and habitat variables.

This study provided clues on the most important habitat features for these various birds. For example it suggests that of the measured habitat variables, the presence of a herb layer was most correlated with numbers of Blackcap. Redstarts by contrast showed a negative correlation with herb cover, reflecting the fact that this species prefers more open, often heavily grazed, woodlands. The numbers of Robin, Great Tit, Blue Tit, Song Thrush and Chaffinch were not statistically correlated with any of the measured habitat variables, probably because these species are quite flexible in their habitat requirements. Such information is invaluable when habitat management programmes are being designed. Any changes in the population of these bird species in the area can also be assessed in terms of changes in the vegetational structure with which they are associated.

Similar studies can be undertaken using transect counts of birds with habitat measured at either regular, randomised or selected positions along the transect (e.g. Hill *et al.* 1990, 1991).

It is important to note that spurious correlations may result when habitat variables are themselves correlated. For example, a dense shrub layer will tend to produce a sparse ground layer and both of these may be statistically significantly correlated with the numbers of a particular bird, although a broad ecological understanding of the bird would suggest that it spends all its time in the shrub layer and never uses the ground. However, on occasion

Figure 11.10

Effects of structural change in vegetation on bird populations

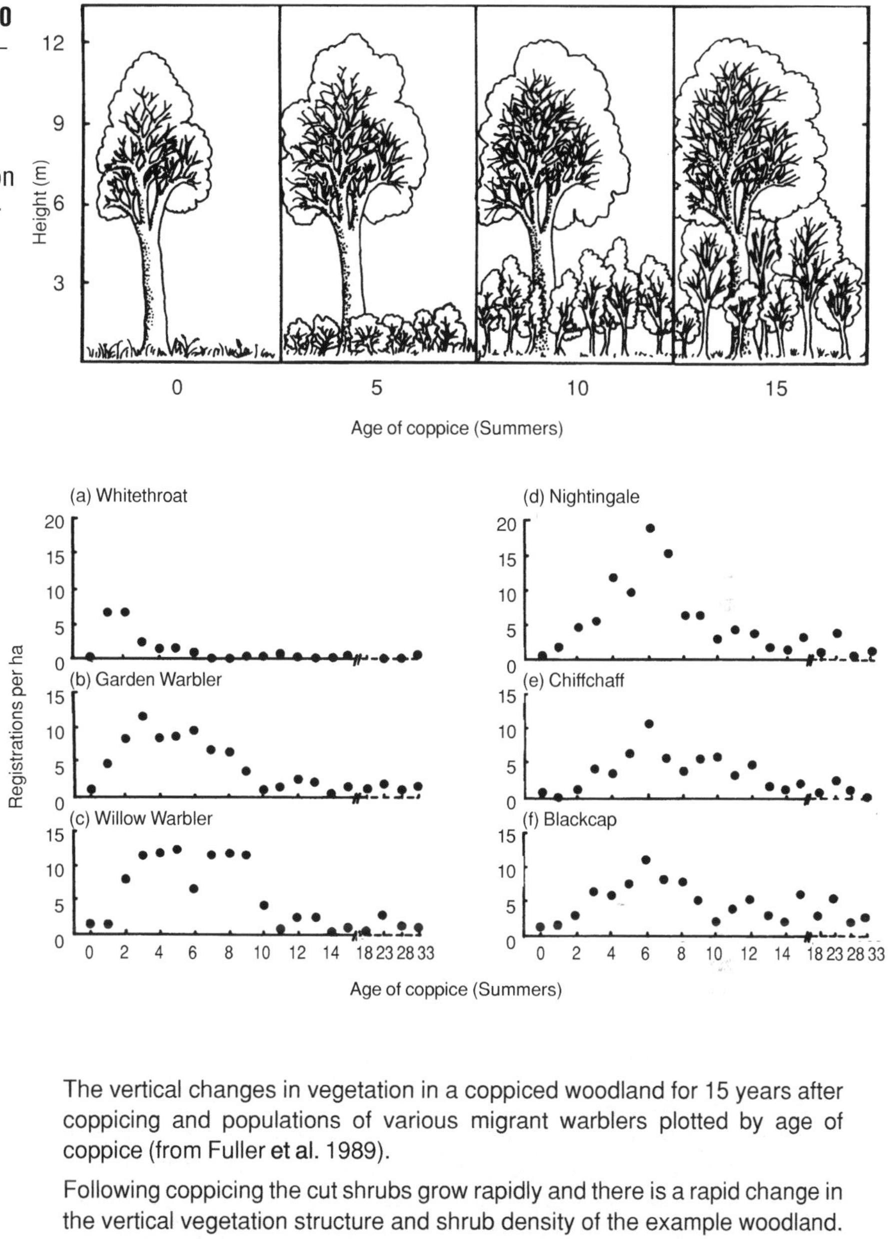

The vertical changes in vegetation in a coppiced woodland for 15 years after coppicing and populations of various migrant warblers plotted by age of coppice (from Fuller et al. 1989).

Following coppicing the cut shrubs grow rapidly and there is a rapid change in the vertical vegetation structure and shrub density of the example woodland. This rapid structural change is reflected in populations of various African migrant warbler species (a–f) which select different structural characteristics (in this case age/density) of coppice. After around 10 years of growth the structure of the coppiced shrub layer is changing much more slowly and the bird community becomes more stable. Populations of the African migrant warbler species fall, and are replaced by resident birds such as tits and thrushes (not shown). The species graphs are presented (a–f) approximately in order of habitat selection with species preferring the youngest coppice coming first.

Table 11.4

Habitat factors making a significant (P < 0.05) contribution to explaining bird numbers, as assessed by multiple regression analysis (from Bibby and Robins 1985)

							Variables							
	1	2	3	4	5	6	7	8	9	10	11	12	13	% Variance explained
Blackcap							+							54.7
Willow Warbler		+	+	+										49.8
Chiffchaff				+										45.7
Wood Warbler					+	+					−			71.0
Goldcrest		−		−										62.9
Pied Flycatcher	−								+	+	−			72.5
Redstart						−	−							59.3
Blackbird		+	+		+									64.3
Willow Tit									−		−		+	60.6
Nuthatch							+					−	+	57.5
Treecreeper	−					−								54.0
Wren										+				61.6
Song Thrush														NS
Robin														NS
Great Tit														NS
Blue Tit														NS
Chaffinch														NS

Key to variables: 1 = birch %, 2 = sessile oak %, 3 = tree species diversity, 4 = canopy cover diversity, 5 = shrub cover, 6 = foliage height diversity, 7 = herb cover, 8 = density trees > 15 m, 9 = % overmature trees, 10 = holes, 11 = hazel %, 12 = bramble %, 13 = scrub height diversity.

The higher the variance explained the better the correlation with that particular variable. Values of over 60% are highly significant.

apparently spurious correlations may prompt a useful re-investigation into the habitat preferences of the bird!

Individual-based studies

As well as relating bird counts to mapped habitat features, and habitat variables measured in sample plots, it is also possible to record habitat variables at the exact location of a singing bird, or one located by radio-telemetry. Such studies can be used to define habitat selection by a bird more precisely, and if habitat variables are recorded at positions with and without a bird, then a habitat preference index can be created (see later).

1. Radio-marked Partridges on farmlands in Britain

Early morning (roosting site) locations of female Red-legged and Grey Partridges with young were assessed by radio-telemetry (Green 1984), and then by visiting the precise location of the radio-fix and finding the roosting area, habitat variables such as crop type and distance from the

Figure 11.11

Radio-
telemetry as
a tool for
studying
habitat pref-
erences of
individual
species

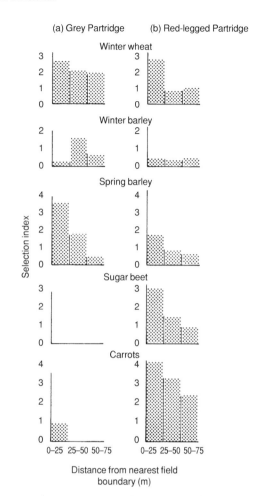

Habitat selection by (a) Grey Partridge and (b) Red-legged Partridge in Britain
is shown (from Green 1984). Histograms show ratios of observed to expected
numbers of radio-locations in five crops and at various distances from the
nearest field boundary. This shows the preference of these two species for
areas adjacent to hedges, and also shows that they prefer different crop types.

nearest hedge were measured. Moreover, droppings were collected and
used to assess what the birds had been feeding on.

After many such sites had been visited and habitat variables collected
and analysed, it was shown that both species roost most often close to
hedges, but that the habitat preference varies between the species. Grey
Partridge preferred cereals, whereas Red-legged Partridge preferred
carrots and sugar beet crops within the study area (Fig. 11.11).

2. Radio-marked Woodcock feeding in woodlands and fields

Radio-telemetry techniques have also been used to locate Woodcock within a study area and assess their habitat preferences. The study plot was first divided into four strata (trees, saplings, shrubs and herbaceous). Radio-tagged Woodcock were then located within these broad habitat divisions by radio-telemetry and details of the habitat were collected by means of a 0.25 m² quadrat at feeding, nesting and randomly located sites. In fact, 30 habitat variables were recorded at each of the 50 feeding locations. By recording habitat variables in areas used by feeding Woodcock and in randomly located quadrats a habitat preference index could be developed, showing that the Woodcock positively selected some of the habitat attributes in the study area (Table 11.5).

Table 11.5

Means and standard deviations of habitat variables that were significantly different (P< 0.05) between Woodcock feeding locations as determined by radio-telemetry and random sites (Hirons and Johnson 1987)

Variable	Woodcock feeding areas Mean	SD	Randomly chosen areas Mean	SD	Significance level
Vegetation structure					
Basal area of trees (m² per ha)	2.8	1.73	4.4	4.20	<0.05
Mean basal area of trees (cm²)	27.8	10.50	37.7	18.50	<0.01
Height of co-dominant vegetation (cm)	8.2	6.96	4.9	6.73	<0.05
Vegetation composition					
Dog's Mercury (% cover)	19.9	13.34	1.3	5.37	<0.001
Beech (% of point quarters)	12.4	25.30	40.6	40.63	<0.05
Oak (% of point quarters)	3.2	7.78	0.3	3.08	<0.05
Ground surface and soil characteristics					
Litter (% cover)	41.5	14.52	62.6	25.00	<0.05
Litter depth (cm)	1.5	1.14	2.5	2.67	<0.05
pH	6.3	0.79	5.2	1.15	<0.001
Earthworm biomass (g per 0.25 m²)	5.62	3.21	3.08	3.27	<0.001
Earthworm numbers (per 0.25 m²)	18.92	1.74	10.41	1.80	<0.001

In this analysis the lower the level of P the higher the significance of the result, hence the most significant results are:

1. the preference for feeding within Dog's Mercury;
2. the preference for feeding in areas with higher soil pH, earthworm numbers and earthworm biomass.

Summary and points to consider

Bird habitats are generally recorded within three broad divisions: (1) mapping methods – producing habitat maps; (2) sample plot methods – measuring habitat variables in representative sample plots; (3) individual-based methods – measuring habitat variables at the known position of a study bird.

1. Mapping methods
Using habitat maps
Base habitat maps are produced from national geographical maps, aerial photographs or satellite images. They are refined by ground survey and, at the highest intensity of effort, habitats can be classified down to the vegetation community level. If registrations of birds are marked on habitat maps they provide basic information on the birds' habitat preferences, but cannot determine which detailed features of the habitat are most important to the birds.

Modelling species distributions
Given prior knowledge of the habitat-preferences of a particular bird species, habitat maps can be used to predict bird distribution and population levels over extensive areas.

2. Measurement of habitat variables in sample plots
Habitat variables can be recorded in sample plots at the position of a point count or along a transect. Measurement of habitat variables of probable importance to the bird, as derived from prior knowledge of its ecology, enables the relative importance of habitat features to be investigated. It is then possible to discover which habitat variables are the most important to an assemblage of birds, or an individual bird species.

3. Measurement of habitat variables at the position of a study bird
Habitat variables can also be measured at the precise position of a breeding or feeding bird (e.g. located by radio-telemetry). A more refined level of knowledge of the important habitat features to the bird should result from this form of study.

Appendix

Scientific names of species mentioned in the text

Mammals

Fox   *Vulpes vulpes*

Plants

Ash   *Fraxinus excelsior*
Beech   *Fagus sylvatica*
Birch   *Betula* spp.
Bramble   *Rubus fruticosus agg.*
Dog's Mercury   *Mercurialis perennis*
Field Maple   *Acer campestre*
Hazel   *Corylus avellana*
Pedunculate Oak   *Quercus robur*
Sessile Oak   *Quercus petraea*

References

Aebischer, N.J. and Robertson, P.A. (1992). Practical aspects of compositional analysis as applied to Pheasant habitat utilisation. In: Priede, I.G. and Swift, S.M. (eds.). *Wildlife Telemetry: Remote Monitoring and Tracking of Animals.* Ellis Harwood, Chichester, UK.

Aitchison, J. (1986). *The Statistical Analysis of Compositional Data.* Chapman and Hall, London.

Alerstam, T. (1990). *Bird Migration.* Cambridge University Press, Cambridge.

Alexander, H.G. (1935). A chart of bird song. *British Birds* **29**, 190–198.

Alexander, M. and Perrins, C.M. (1980). An estimate of the numbers of shearwaters on the Neck, Skomer, 1978. *Nature in Wales* **17**, 43–46.

Alldredge, J.R. and Ratti, J.T. (1986). Comparison of some statistical techniques for analysis of resource selection. *Journal of Wildlife Management* **50**, 157–165.

Andreev, A. (1988). The ten year cycle of the Willow Grouse of lower Kolyma. *Oecologia* **76**, 261–267.

Anker-Nilssen, T. and Røstad, O.W. (1993). Census and monitoring of puffins *Fratercula arctica* on Rost, N. Norway, 1979–1988. *Ornis Scandinavica* **24**, 1–9.

Arbib, R.S. (1981). The Christmas Bird Count: constructing an 'ideal model'. *Studies in Avian Biology* **6**, 30–33.

Avery, M.I. (1989). The effects of upland afforestation on some birds of the adjacent moorlands. *Journal of Applied Ecology* **26**, 957–967.

Avery, M.I. and Haines-Young, R.H. (1990). Population estimates derived from remotely-sensed imagery for *Calidris alpina* in the Flow Country of Caithness and Sutherland. *Nature* **344**, 860–862.

Baillie, S.R. (1990). Integrated population monitoring of breeding birds in Britain and Ireland. *Ibis* **132**, 151–166.

Barnes, R.F.W. (1987). Long-term declines of Red Grouse in Scotland. *Journal of Applied Ecology* **24**, 735–741.

Barrett, J. and Barrett, C. (1984). Aspects of censusing breeding Lapwings. *Wader Study Group Bulletin* **42**, 45–47.

Bart, J. and Klosiewski, S.P. (1989). Use of presence–absence to measure changes in avian density. *Journal of Wildlife Management* **53**, 847–852.

Bayliss, P. (1989). Population dynamics of Magpie Geese in relation to rainfall and density: implications for harvest models in a fluctuating environment. *Journal of Applied Ecology* **26**, 913–924.

Bednarz, J.C. and Kerlinger, P. (1989). Monitoring hawk populations by counting migrants. pp. 328–342. In: *Proceedings of the Northeast Raptor Management Symposium and Workshop.* National Wildlife Federation, Washington D.C.

Begon, M. and Mortimer, M. (1986). *Population Ecology. A Unified Study of Animals and Plants.* Blackwell Scientific Publications, Oxford.

Belant, J.L. and Seamans, T.W. (1993). Evaluation of dyes and techniques to color-mark incubating Herring Gulls. *Journal of Field Ornithology* **64**, 440–451.

Bell, H.L. (1982). A bird community of lowland rainforest in New Guinea. I. Composition and density of the avifauna. *Emu* **82**, 24–41.

Bellrose, F.C. (1976). *Ducks, Geese and Swans of North America* (2nd edn). Stackpole Books, Harrisburg.

Bergan, J.F. and Smith, L.M. (1989). Differential habitat use by diving ducks wintering in South Carolina. *Journal of Wildlife Management* **53**, 1117–1126.

Bergan, J.F., Smith, L.M. and Mayer, J.J. (1989). Time–activity budgets of diving ducks wintering in South Carolina. *Journal of Wildlife Management* **53**, 769–776.

Berthold, P., Fliege, G., Querner, U. and Winkler, H. (1986). Die Bestandsentwicklung von Kleinvögeln in Mitteleuropa: Analyse von Fangzahlen. *Journal für Ornithologie* **127**, 377–439.

Berthold, P., Kaiser, A., Querner, V. and Schlenker, R. (1993). Analysis of trapping figures at Mettnau station S. Germany, with respect to the population development of small birds: a 20 year summary [in German with English summary]. *Journal of Ornithology* **134**, 283–299.

Bezzel, E. (1985, 1993). *Kompendium der Vögel Mitteleuropas.* 2 Bände (Nonpasseriformes und Passers). Aula-Verlag, Wiesbaden.

Bibby, C.J. (1973).The Red-backed Shrike: a vanishing British species. *Bird Study* **20**, 103–110.

Bibby, C.J. (1978). A heathland bird census. *Bird Study* **25**, 87–96.

Bibby, C.J. (1989). A survey of breeding Wood Warblers *Phylloscopus sibilatrix* in Britain, 1984–1985. *Bird Study* **36**, 56–72.

Bibby, C.J. and Buckland, S.T. (1987). Bias of bird census results due to detectability varying with habitat. *Acta Oecologica–Oecologica Generalis* **8**, 103–112.

Bibby, C.J., Burgess, N.D. and Hill, D.A. (1992). *Bird Census Techniques.* Academic Press, London.

Bibby, C.J., Burgess, N.D. and Hill, D.A. (German translation by Bauer, H-G.). (1995). *Methoden der Feldornithologie: bestanderfassung in der praxis.* Neumann, Leipzig.

Bibby, C.J. and Charlton, T.D. (1991). Observations on the San Miguel Bullfinch. *Açoreana* **7**, 297–304.

Bibby, C.J. and Hill, D.A (1987). Status of the Fuerteventura Stonechat *Saxicola dacotiae. Ibis* **129**, 491–498.

Bibby, C. J., Jones, M. and Marsden, S. (1998) *Expedition Field Techniques: Bird Surveys.* With BirdLife International and the Expedition Advisory Centre. EAC, Royal Geographical Society (with the Institute of British Geographers), London.

Bibby, C.J., Phillips, B.N. and Seddon, A.J. (1985). Birds of restocked conifer plantations in Wales. *Journal of Applied Ecology* **22**, 619–633.

Bibby, C.J. and Robins, M. (1985). An exploratory analysis of species and community relationships with habitat in western oak woods, pp. 255–265. In: Taylor, K., Fuller, R.J. and Lack, P.C. (eds.). *Bird Census and Atlas Studies. Proceedings, III International Conference on Bird Census and Atlas Work.* BTO, Tring, Herts.

Bibby, C.J. and Tubbs, C.R. (1975). Status and conservation of the Dartford Warbler in England. *British Birds* **68**, 177–195.

Bierregaard, R.O. and Lovejoy, T.E. (1988). Birds in Amazonian forest fragments: Effects of insularisation. pp. 1564–1579. In: Ouellet, H. (ed.). *Acta XIX Congressus Internationalis Ornithologici, Vol II.* University of Ottawa Press, Canada.

Bierregaard, R.O. and Lovejoy, T.E. (1989). Effects of forest fragmentation on Amazonian understorey bird communities. *Acta Amazonica* **19**, 215–241.

Bildstein, K.L. and Zalles, J.I. (eds.) (1995). *Raptor Migration Watch-site Manual.* Hawk Mountain Sanctuary Association, Kempton, Pennsylvania, USA.

Birkhead, T.R. and Nettleship, D.N. (1980). Census methods for Murres *Uria* species: a unified approach. *Occasional Papers of the Canadian Wildlife Service* **43**.

Bock, C.E. and Root, T.L. (1981). The Christmas Bird Count and avian ecology. *Studies in Avian Biology* **6**, 17–23.

Boddy, M. (1993). Whitethroat population studies during 1981–91 at a breeding site on the Lincolnshire coast. *Ringing and Migration* **14**, 73–83.

Boinski, S. and Scott, P.E. (1988). Association of birds with monkeys in Costa Rica. *Biotropica* **20**, 136–143.

Briggs, K.T., Tyler, W.B. and Lewis, D.B. (1985). Comparison of ship and aerial surveys of birds at sea. *Journal of Wildlife Management* **49**, 405–411.

British Birds (1984). *The 'British Birds' List of Birds of the Western Palearctic.* British Birds, Bedford.

Brosset, A. (1990). A long term study of the rain forest birds in M'Passa (Gabon). In: Keast, A. (ed.). *Biogeography and Ecology of Forest Birds.* SPB Academic, The Hague, The Netherlands.

Brosset, A. and Erard, C. (1986). *Les Oiseaux des Régions Forestières du Nord-est du Gabon. Volume 1. Écologie et Comportement des Espèces.* Société Naturale de Protection de la Nature, Paris, France.

Brown, A.F. and Shepherd, K.B. (1993). A method for censusing upland breeding waders. *Bird Study* **40**, 189–195.

Brown, L.H., Fry, C.H., Keith, S., Newnam, K. and Urban, E.K. (eds.) (1982 onwards). *The Birds of Africa. Vols I–III.* Academic Press, London.

Bruderer, B. (1992). Radar studies on bird migration in the Negev. *Ostrich* **65**, 204–212.

BTO (1984). *Ringers Manual.* BTO, Tring, Herts.

BTO (1989). *Instruction to Counters: Breeding Waders of Wet Grasslands Survey.* BTO, Tring, Herts.

BTO (1991). *Guidelines and Site Record Form for the Fourth BTO Peregrine Survey 1991.* BTO, Thetford.

Buckland, S.T. (1984). Monte Carlo confidence intervals. *Biometrics* **40**, 811–817.

Buckland, S.T., Anderson, D.R., Burnham, K.P. and Laake, J.L. (1993). *Distance Sampling. Estimating Abundance of Biological Populations.* Chapman and Hall, London.

Buckland, S.T., Shewry, M.C. and Shaw, P. (1998). *The Application of Distance Sampling to Monitoring Selected Species on Designated Sites.* Scottish Natural Heritage internal report.

Buker, J.B. and Groen, N.M. (1989). Distribution of Black-tailed Godwits *L. limosa* in different grassland types during the breeding season. *Limosa* **62**, 183–190.

Bullock, I.D. and Gomersall, C.H. (1981). The breeding populations of terns in Orkney and Shetland in 1980. *Bird Study* **28**, 187–200.

Bundy, G. (1978). Breeding Red-throated Divers in Shetland. *British Birds* **71**, 199–208.

Bunn, D.S., Warburton, A.B. and Wilson, R.D.S. (1982). *The Barn Owl.* T & AD Poyser, Calton.

Burger, A.E. (1997). Behaviour and numbers of Marbled Murrelets measured with radar. *Journal of Field Ornithology,* **68**, 208–223.

Burnham, K.P. and Anderson, D.R. (1984). The need for distance data in transect counts. *Journal of Wildlife Management* **48**, 1248–1254.

Busse, P. (1990). Studies of long-term population dynamics based on ringing data. *Ring* **13**, 221–234.

Cadbury, C.J. (1980). The status and habitats of the Corncrake in Britain 1978–79. *Bird Study* **27**, 203–218.

Cadbury, C.J. (1981). Nightjar census methods. *Bird Study* **28**, 1–4.

Campbell, L.H. and Talbot, T.R. (1987). Breeding status of Black-throated Divers in Scotland. *British Birds* **80**, 1–8.

Casagrande, D.G. and Beissinger, S.R. (1997). Evaluation of four methods for estimating parrot population size. *Condor* **99**, 445–457.

Cavanagh, P.M. and Griffin, C.R. (1993). Suitability of Velcro leg tags for marking Herring and Great Black-backed Gull chicks. *Journal of Field Ornithology* **64**, 195–198.

Cayford, J.T. and Walker, F. (1991). Counts of male Black Grouse *Tetrao tetrix* in north Wales. *Bird Study* **38**, 80–86.

Clark, N.A. (1990). *Distribution Studies of Waders and Shelduck in the Severn Estuary.* Report to UK Department of Energy's Renewable Energy Research and Development Programme (ETSU TID 4076), London.

Clark, W.S. and Wheeler, B.K. (1987). *A Field Guide to the Hawks of North America.* Houghton Mifflin Co., Boston.

Clarke, R. and Watson, D. (1990). The Hen Harrier *Circus cyaneus* winter roost survey in Britain and Ireland. *Bird Study* **37**, 84–100.

Clobert, J., Lebreton, J.D. and Allaine, D. (1987). A general approach to survival rate estimation by recaptures or resightings of marked birds. *Ardea* **75**, 133–142.

Cody, M.L. (1985). *Habitat Selection in Birds.* Academic Press, London.

Collar, N.J. and Andrew, P. (1988). *Birds to Watch: the ICBP World Checklist of Threatened Birds.* ICBP Technical Publication No. 8. ICBP, Cambridge.

Collar, N.J., Crosby, M.J. and Stattersfield, A.J. (1994). *Birds to Watch 2: The World List of Threatened Birds.* BirdLife International, Cambridge. BirdLife Conservation Series No. 4.

Colwell, R.K. and Coddington, J.A. (1994). Estimating terrestrial biodiversity through extrapolation. *Phil. Trans. R. Soc. Lond.* B. **345**, 101–118.

Conner, R.N. (1990). The effect of observer variability on the MacArthur foliage density estimates. *Wilson Bulletin* **102**, 341–343.

Conroy, M.J., Hines, J.E. and Williams, B.K. (1989). Procedures for the analysis of bird-recovery data and user instructions for programme MULT. *Resource publication–US Fish and Wildlife Service* **175**, 1–61.

Cooch, E.G., Lark, D.B., Rockwell, R.F. and Cooke, F. (1989). Long term decline in fecundity in a Snow Goose population: evidence for density dependence. *Journal of Animal Ecology* **58**, 711–726.

Cooper, B.A., Day, R.H., Ritchie, R.J. and Cranor, C.L. (1991). An improved marine radar system for studies of bird migration. *Journal of Field Ornithology* **62**, 367–377.

Cormack, R.M. (1968). The statistics of capture–recapture methods. *Ocean Marine Biology Annual Review* **6**, 455–506.

Cormack, R.M. (1979). Models for capture–recapture, pp. 217–255. In: Cormack, R.M., Patil, G.P. and Robson, D.S. (eds.). *Sampling Biological Populations. Statistical Ecology Series. Vol 5.* International Co-op Publishing House, Fairland, Maryland, USA.

Cramp. S. and Simmons, K.E.L. (eds.) (1977 onwards). *The Birds of the Western Palearctic. Vols I–V.* Academic Press, London.

Cranswick, P.A., Kirby, J.S., Salmon, D.G., Atkinson-Willes, G.L., Pollitt, M.S. and Owen, M. (1996). A history of wildfowl counts by The Wildfowl and Wetlands Trust. *Wildfowl* **47**, 216–226.

Cranswick, P.A., Waters, R.J., Musgrove, A.J. and Pollitt, M.S. (1997). *The Wetland Bird Survey 1995–96: Wildfowl and Wader Counts.* BTO/WWT/RSPB/JNCC, Slimbridge.

Crooke, C., Dennis, R., Harvey, M. and Summers, R. (1992). Population size and breeding success of Slavonian Grebes in Scotland. In *Britain's Birds in 1990–1: The Conservation and Monitoring Review.* British Trust for Ornithology, Thetford.

Danielsen, F. and Heegaard, M. (1995). Impact of logging and plantation development on species diversity: a case study from Sumatra. pp. 73–92. In: Sandbukt, Ø. (ed.) *Management of Tropical Forests: Towards an Integrated Perspective.* Centre for Development and the Environment, University of Oslo, Oslo.

Davis, S.J.J.F. (ed.) (1984). *Methods of Censusing Birds in Australia.* Report No. 7. Royal Australian Ornithological Union, Victoria.

Day, J. (1988). Marsh Harriers in Britain. *RSPB Conservation Review* **2**, 17–19.

del Nevo, A.J. (1990). *Reproductive and Feeding Ecology of Common Guillemots* (Uria aalge) *on Fair Isle, Shetland.* Ph.D. thesis. University of Sheffield.

DeSante, D.F. (1981). Censusing technique in a California coastal scrub breeding bird community. *Studies in Avian Biology* **6**, 177–186.

DeSante, D.F. (1986). A field test of the variable circular-plot censusing method in a Sierran subalpine forest habitat. *Condor* **88**, 129–142.

Diamond, A.W., Gaston, A.J. and Brown, R.G.B. (1986). Converting PRIOP counts of seabirds at sea to absolute densities. *Progress Notes of Canadian Wildlife Service* **164**, 1–21.

Dixon, K.R. and Chapman, J.A. (1980). Harmonic mean measure of animal activity areas. *Ecology* **61**, 1040–1044.

Dodman, T. and Taylor, V. (1995). *African Waterfowl Census. Les Dénombrements Internationaux d'oiseau d'eau en Afrique, 1995.* International Waterfowl Research Bureau, Slimbridge.

Drennan, S.R. (1981). The Christmas Bird Count: an overlooked and underused sample. *Studies in Avian Biology* **6**, 24–29.

Duebbert, H.F. and Lokemoen, J.T. (1976). Duck nesting in fields of undisturbed grass–legume cover. *Journal of Wildlife Management* **40**, 39–49.

du Feu, C., Hounsome, M. and Spence, I. (1983). A single-session mark/recapture method of population estimation. *Ringing and Migration* **4**, 211–226.

du Feu, C. and McMeeking, J. (1991). Does constant effort netting estimate juvenile abundance? *Ringing and Migration* **12**, 118–123.

Dunn, E.H., Cadman, M. and Falls, J.B. (eds.) (1995). *Monitoring Bird Populations: the Canadian Experience.* Occasional Paper No. 95 of the Canadian Wildlife Service, Ottawa.

Dunn, E.H., Larivee, J. and Cyr, A. (1996). Can checklist programs be used to monitor populations of birds recorded during the migration season? *Wilson Bulletin* **108**, 540–549.

Dunne, P., Keller, D. and Kochenberger, R. (1984). *Hawk Watch: A Guide for Beginners.* Cape

May Bird Observatory, Cape May Point.

Dunne, P., Sibley, D.A. and Sutton, C.C. (1988). *Hawks in Flight, the Flight Identification of North American Migrant Raptors*. Houghton Mifflin Co., Boston.

Dunnet, G.M., Ollason, J.C. and Anderson, A. (1979). A 28-year study of breeding Fulmars *Fulmarus glacialis* in Orkney. *Ibis* **121**, 293–300.

Durinck, J., Skov, H., Jensen, F.P. and Phil, S. (1994). *Important Marine Areas for Wintering Birds in the Baltic Sea*. EU DG XI research contract no. 224/90-09-01, Ornis Consult Report, Copenhagen.

Eastwood, E. (1967). *Radar Ornithology*. Methuen, London.

Efron, B. (1982). *The Jackknife, the Bootstrap and Other Resampling Methods*. Society for Industrial and Applied Mathematics, Philadelphia.

Ehrlich, D. and Lambin, E.F. (1996). Broad scale land-cover classification and interannual climate variability. *International Journal of Remote Sensing* **17**, 845–862.

Ellenberg, H. (1985). How to use species area relationships to compare grid-mapping results from different grid sizes, pp. 321–329. In: Taylor, K., Fuller, R.J. and Lack, P. (eds.). *Bird Census and Atlas Studies. Proceedings VIII International Conference on Bird Census and Atlas Work*. BTO, Tring, Herts.

Ellis, D.H., Glinski, R.L. and Smith, D.G. (1990). Raptor road surveys in south America. *Journal of Raptor Research* **24**, 98–106.

Enemar, A. (1959). On the determination of the size and composition of a passerine bird population during the breeding season. *Vår Fågelvärld Supplement* **2**, 1–114.

Engstrom, R.T. (1990). Evaluation of the Colonial Bird Register. pp. 26–30. In: Sauer, J.R. and Droege, S. (eds.). *Survey Designs and Statistical Methods for the Estimation of Avian Population Trends*. US Fish and Wildlife Service Biological Report **90**(1).

Engstrom, R.T. and James, F.C. (1984). An evaluation of methods used in the Breeding Bird Census. *American Birds* **28**, 19–23.

Erdelen, M. (1984). Bird communities and vegetation structure I. Correlations and comparisons of simple diversity indices. *Oecologia* **61**, 277–284.

Evans, P.G.H. (1980). *Auk Censusing Manual*. British Seabird Group, Tring, Herts.

Evans, P.G.H. (1986). Monitoring seabirds in the North Atlantic. *NATO ASI Series* **G12**, 179–206.

Evans, W.R. (1994). Nocturnal flight call of Bicknell's Thrush. *The Wilson Bulletin* **106**, 55–61.

Everett, M.J. (1982). Breeding Great and Arctic Skuas in Scotland in 1974–75. *Seabird Report* **6**, 50–58.

Ewins, P.J. (1985). Colony attendance and censusing of Black Guillemots *Cepphus grylle* in Shetland. *Bird Study* **32**, 176–185.

Fjeldså, J. (1999). The impact of human forest disturbance on the endemic avifauna of the Udzungwa Mountains, Tanzania. *Bird Conservation International* **9**, 47–62.

Fjeldså, J. and Rabøl, J. (1995). Variation in avian communities between isolated units of the Eastern Arc Montane forests, Tanzania. *Gerfaut* **85**, 3–18.

Fowler, J. and Cohen, L. (1986). *Statistics for Ornithologists*. BTO, Tring, Herts.

Fowler, J. and Cohen, L. (1990). *Basic Statistics for Field Biologists*. BTO, Tring, Herts.

Fuller, M.R. and Mosher, J.A. (1981). Methods of detecting and counting raptors. *Studies in Avian Biology* **6**, 235–246.

Fuller, M.R. and Titus, K. (1990). pp. 41–46. In: Sauer, J.R. and Droege, S. (eds.). *Survey Designs and Statistical Methods for the Estimation of Avian Population Trends*. US Fish and Wildlife Service Biological Report **90**(1).

Fuller, R.J. (1982). *Bird Habitats in Britain*. T & AD Poyser, Calton.

Fuller, R.J., Green, G.H. and Pienkowski, M.W. (1983). Field observations on methods used to count waders breeding at high density in the Outer Hebrides, Scotland. *Wader Study Group Bulletin* **39**, 27–29.

Fuller, R.J., Gregory, R.D., Gibbons, D.W., Marchant, J.H., Wilson, J.D., Baillie, S.R., Carter, N. (1995). Population declines and range contractions among lowland farmland birds in Britain. *Conservation Biology* **9**, 1425–1441.

Fuller, R.J. and Langslow, D.R. (1984). Estimating numbers of birds by point counts: how long should counts last? *Bird Study* **31**, 195–202.

Fuller, R.J. and Marchant, J.H. (1985). Species-specific problems of cluster analysis in British mapping censuses, pp. 83–86. In: Taylor, K., Fuller, R.J. and Lack, P.C. (eds.). *Bird Census and Atlas Studies. Proceedings VIII International Conference on Bird Census and Atlas Work.* BTO, Tring, Herts.

Fuller, R.J., Marchant, J.H. and Morgan, R.A. (1985). How representative of agricultural practice in Britain are Common Bird Census farmland plots? *Bird Study* **32**, 56–70.

Fuller, R.J. and Moreton, B.D. (1987). Breeding bird populations of Kentish sweet chestnut (*Castanea sativa*) coppice in relation to age and structure of the coppice. *Journal of Applied Ecology* **24**, 13–27.

Fuller, R.J., Reed, T.M., Buxton, N.E., Webb, A., Williams, T.D. and Pienkowski, M.W. (1986). Populations of breeding waders Charadrii and their habitats on the crofting lands of the Outer Hebrides, Scotland. *Biological Conservation* **37**, 333–361.

Fuller, R.J., Stuttard, P. and Ray, C.M. (1989). The distribution of breeding songbirds within mixed coppiced woodland in Kent, England, in relation to vegetation age and structure. *Annales Zoologici Fennici* **26**, 265–275.

Furness, R.W. (1982). Methods used to census skua colonies. *Seabird Report* **6**, 44–47.

Furness, R.W. and Greenwood, J.J.D. (eds.) (1993). *Birds as Monitors of Environmental Change.* Chapman and Hall, London.

Gaston, A.J., Collins, B.T. and Diamond, A.W. (1987). Estimating densities of birds at sea and the proportion in flight from counts made on transects of indefinite width. *Canadian Wildlife Service Occasional Paper* **59**, 1–14.

Gauch, H.G. (1982). *The Use of Multivariate Analysis in Community Ecology.* Cambridge University Press, Cambridge.

Geissler, P.H. and Noon, B.R. (1981). Estimates of avian population trends from the North American Breeding Bird Survey. *Studies in Avian Biology* **6**, 45–61.

Gibbons, D., Avery, M., Baillie, S., Gregory, R., Kirby, J., Porter, R., Tucker, G. and Williams, G. (1996). Bird Species of Conservation Concern in the United Kingdom, Channel Islands and Isle of Man: revising the Red Data List. In: *RSPB Conservation Review No. 10* (ed. C.J. Cadbury). RSPB, Sandy. 7–18.

Gibbons, D.W., Hill, D. and Sutherland, W.J. (1996). Birds. pp. 227–259. In: Sutherland, W.J. (ed.). *Ecological Census Techniques: A Handbook.* Cambridge University Press, Cambridge.

Gibbons, D.W., Reid, J.B. and Chapman, R.A. (1993). *The New Atlas of Breeding Birds in Britain and Ireland: 1988–1991.* T & AD Poyser, London.

Gilbert, G., Gibbons, D.W. and Evans, J. (1998) *Bird Monitoring Methods: A Manual of Techniques for Key UK Species.* RSPB/BTO/JNCC/WWT/ITE/The Seabird Group RSPB/BTO, Sandy.

Gilbert, G., McGregor, P.K. and Tyler, G. (1994). Vocal individuality as a census tool: practical considerations illustrated by a study of two rare species. *Journal of Field Ornithology* **65**, 335–348.

Ginsból, B. (1984). *Collins Guide to the Birds of Prey of Britain and Europe.* Collins, London.

Gochfield, M. (1980). Mechanisms and adaptive values of reproductive synchrony in colonial seabirds. pp. 207–270. In: Burger, J., Olla, B.L. and Winn, H.E. (eds.). *Behaviour of Marine Animals. Vol. 4. Marine Birds.* Plenum Press, New York.

Goldsmith, F.B. (ed.) (1991). *Monitoring for Conservation and Ecology.* Chapman and Hall, London.

Goss-Custard, J.D. and Durell, S.E.A. le V. Dit (1990). Bird behaviour and environmental planning: approaches in the study of wader populations. *Ibis* **132**, 273–289.

Gotelli, N.J. and Graves, G.R. (1995). *Null Models in Ecology.* Smithsonian Institution Press, Washington DC.

Green, R.E. (1984). The feeding ecology and survival of partridge chicks (*Alectoris rufa* and *Perdix perdix*) on arable farmland in East Anglia. *Journal of Animal Ecology* **21**, 817–830.

Green, R.E. (1985a). Estimating the abundance of breeding Snipe. *Bird Study* **32**, 141–149.

Green, R.E. (1985b). *The Management of Lowland Wet Grasslands for Breeding Waders.* RSPB, Sandy, Beds.

Green, R.E. (1988). Effects of environmental factors on the timing and success of breeding of Common Snipe *Gallinago gallinago* (Aves: Scolopacidae). *Journal of Applied Ecology* **25**, 79–93.

Green, R.E. (1995). The decline of the Corncrake *Crex crex* in Britain continues. *Bird Study* **42**, 66–75.

Green, R.E. and Hirons, G.J.M. (1988). Effects of nest failure and spread of laying on counts of breeding birds. *Ornis Scandinavica* **19**, 76–78.

Green, R.E. and Hirons, G.J.M. (1990). The relevance of population studies to the conservation of threatened birds, pp. 595–631. In: Perrins, C.M., Lebreton, J.D. and Hirons, G.J.M. (eds.). *Bird Population Studies*. Oxford University Press, Oxford.

Gregory, R.D. and Baillie, S.R. (1998). Large-scale habitat use of some declining British birds. *Journal of Applied Ecology* **35**, 785–799.

Gregory, R.D., Bashford, R.I., Balmer, D.B., Marchant, J.H., Wilson, A.M. and Baillie, S.R. (1996). *The Breeding Bird Survey 1994–1995*. British Trust for Ornithology, Thetford.

Gregory, R.D., Bashford, R.I., Balmer, D.B., Marchant, J.H., Wilson, A.M. and Baillie, S.R. (1997). *The Breeding Bird Survey 1995–1996*. British Trust for Ornithology, Thetford.

Gribble, F.C. (1983). Nightjars in Britain and Ireland in 1981. *Bird Study* **30**, 165–176.

Hagemeijer, E.J.M. and Blair, M.J. (eds.) (1997). *The EBCC Atlas of European Breeding Birds: Their Distribution and Abundance*. T & A D Poyser, London.

Haila, Y. and Kuuesla, S. (1982). Efficiency of one-visit censuses of bird communities breeding on small islands. *Ornis Scandinavica* **13**, 17–24.

Hairston, N.G. (1989). *Ecological Experiments: Purpose, Design and Execution*. Cambridge University Press, Cambridge.

Hancock, M.H., Gibbons, D.W. and Thompson, P.S. (1997). The status of breeding Greenshank *Tringa nebularia* in the United Kingdom in 1995. *Bird Study* **44**, 290–302.

Hanowski, J.M. and Niemi, G.J. (1995). A comparison of on- and off-road bird counts: do you need to go off road to count birds accurately? *Journal of Field Ornithology* **66**, 469–483.

Hanssen, O.J. (1982). Evaluation of some methods for censusing larid populations. *Ornis Scandinavica* **13**, 183–188.

Harris, M.P. (1983). *The Puffin*. T & AD Poyser, Calton.

Harris, M.P. (1987). A low-input method of monitoring Kittiwake *Rissa tridactyla* breeding success. *Biological Conservation* **41**, 1–10.

Harris, M.P. (1988). Variation in the correction factor used for converting counts of individual Guillemots *Uria aalge* into breeding pairs. *Ibis* **131**, 85–93.

Harris, M.P. (1989). *Development of Monitoring of Seabird Populations and Performance*. Institute of Terrestrial Ecology: final report to Nature Conservancy Council, Peterborough.

Harris, M.P. and Forbes, R. (1987). The effect of date on counts of nests of Shags *Phalacrocorax aristotelis*. *Bird Study* **34**, 187–190.

Harris, M.P. and Lloyd, C.S. (1977). Variations in counts of seabirds from photographs. *British Birds* **70**, 200–205.

Harris, M.P. and Murray, S. (1981). Monitoring of Puffin numbers at Scottish colonies. *Bird Study* **28**, 15–20.

Harris, M.P. and Rothery, P. (1988). Monitoring of Puffin burrows on Dun, St Kilda, 1977–1987. *Bird Study* **35**, 97–99.

Harrison, J.A., Allan, D.G., Underhill, L.G., Tree, A.J. and Brown, C.J. (eds.) (1997). *The Atlas of Southern African Birds*. SABAP. University of Cape Town, Cape Town.

Hatch, S.A. and Hatch, M.A. (1989). Attendance patterns of Murres at breeding sites: implications for monitoring. *Journal of Wildlife Management* **53**, 483–493.

Haug, E.A. and Diduik, A.B. (1993). Use of recorded calls to detect burrowing owls. *Journal of Field Ornithology* **64**, 188–194.

Haugh, J.R. (1972). A study of hawk migration in eastern North America. *Search* **2**, 1–60.

Hestbeck, J.B. and Malecki, R.A. (1989). Mark–resight estimate of Canada Goose midwinter numbers. *Journal of Wildlife Management* **53**, 749–752.

Heubeck, M., Richardson, M.G. and Dore, C.P. (1986). Monitoring numbers of Kittiwakes *Rissa tridactyla* in Shetland. *Seabird* **9**, 32–42.

Hildén, O. (1986). Long-term trends in the Finnish bird fauna: methods of study and some results. *Vår Fågelvärld Supplement* **11**, 61–69.

Hildén, O. (1987). Finnish winter bird censuses: long-term trends in 1956–84. *Acta Oecologica–Oecologica Generalis* **8**, 157–168.

Hill, D.A. (1982). *The Comparative Population Ecology of Mallard and Tufted Duck*. D.Phil. thesis, University of Oxford.

Hill, D.A. (1984a). Factors affecting nest success in the Mallard and Tufted duck. *Ornis Scandinavica* **15**, 115–122.

Hill, D.A. (1984b). Clutch predation in relation to nest density in Mallard and Tufted Duck. *Wildfowl* **35**, 151–156.

Hill, D.A. (1988). Population dynamics of the avocet (*Recurvirostra avosetta*) breeding in Britain. *Journal of Animal Ecology* **57**, 669–683.

Hill, D.A., Lambton, S.J., Proctor, I. and Bullock, I. (1991). Winter bird communities in woodland in The Forest of Dean, England, and some implications of livestock grazing. *Bird Study* **38**, 57–71.

Hill, D.A. and Robertson, P.A. (1988). *The Pheasant: Ecology, Management and Conservation*. Blackwell Scientific Publications, Oxford.

Hill, D.A., Taylor, S., Thaxton, R., Amphlet, A. and Horn, W. (1990). Breeding bird communities of native pine forest, Scotland. *Bird Study* **37**, 133–141.

Hirons, G.J.M. (1980). The significance of roding by Woodcock *Scolopax rusticola*: an alternative explanation based on observations of marked birds. *Ibis* **22**, 350–354.

Hirons, G.J.M. and Johnson, T.H. (1987). A quantitative analysis of habitat preferences of Woodcock *Scolopax rusticola* in the breeding season. *Ibis* **129**, 371–382.

Holloway, S. (1996). *The historical atlas of breeding birds in Britain and Ireland 1875–1900*. T & AD Poyser, London.

Horne, J. and Short, J. (1988). A note on the sightability of Emus during an aerial survey. *Australian Wildlife Research* **15**, 647–649.

Housden, S., Thomas, G.T., Bibby, C.J. and Porter, R. (1991). Towards a habitat conservation strategy for bird habitats in Britain. *RSPB Conservation Review* **5**, 9–16.

Howe, M.A. (1990). Methodology of the International Shorebird Survey and Constraints on Trend Analysis. pp. 23–26. In: Sauer, J.R. and Droege, S. (eds.). *Survey Designs and Statistical Methods for the Estimation of Avian Population Trends*. US Fish and Wildlife Service Biological Report **90**(1).

Howe, M.A., Geissler, P.H. and Harrington, B.A. (1989). Population trends of North American shorebirds based on the International Shorebird Survey. *Biological Conservation* **49**, 185–199.

Howes, J. and Bakewell, D. (1989). *Shorebird Studies Manual*. Asian Wetland Bureau Publication No. 55.

Howes, J.R. (1987). *Rapid Assessment Techniques for Coastal Wetland Evaluation. Results of a Workshop held in Selangor, West Malaysia. 1–7 March 1987*. INTERWADER publication no. 24, Kuala Lumpur.

Hudson, P. (1986). *Red Grouse: The Biology and Management of a Wild Gamebird*. The Game Conservancy Trust, Fordingbridge.

Hudson, P. and Rands, M. (1988). *Ecology and Management of Gamebirds*. Blackwell Scientific Publications, Oxford.

Hughes, S.W.M., Bacon, P. and Flegg, J.J.M. (1979). The 1975 census of the Great Crested Grebe in Britain. *Bird Study* **26**, 213–226.

Hunter, I., Croxall, J.P. and Prince, P.A. (1982). The distribution and abundance of burrowing seabirds (Procellariiformes) at Bird Island, South Georgia: I. Introduction and methods. *British Antarctic Survey Bulletin* **56**, 49–67.

Hutto, R.L., Pletschet, S.M. and Hendricks, P. (1986) A fixed-radius point count method for nonbreeding and breeding-season use. *Auk* **103**, 593–602.

Iman, R.L. and Davenport, J.M. (1980). Approximations to the critical region of the Friedman statistic. *Community Statistics* **A9**, 571–595.

International Bird Census Committee (1969). Recommendations for an international standard for a mapping method in bird census work. *Bird Study* **16**, 248–255.

Isaksen, K. and Bakken, V. (1995). Estimation of the breeding density of Little Auks (*Alle alle*). Pp. 37–48. In: Isaksen, K. and Bakken, V. (eds.). *Seabirds in the Northern Barents Sea. Source data for the Impact Assessment of the Effects of Oil Drilling Activity*. Norsk Polarinstitut, Oslo.

Ivlev, V.F. (1961). *Experimental Ecology of the Feeding of Fishes*. Yale University Press, New Haven, Connecticut.

Jackson, J.E. (1985). An evaluation of aerial survey techniques for red-cockaded woodpeckers. *Journal of Wildlife Management* **49**, 1083–1088.

Jacobs, J. (1974). Quantitative measurement of food selection. A modification of the forage ratio and Ivlev's electivity index. *Oecologia* **14**, 413–417.

James, D.A. (1992). Measuring shrubland vegetational structure using avian habitats as an example. *Proceedings Arkansas Academy of Science* **46**, 46–48.

James, F.C. and McCulloch, C.E. (1985). Data analysis and the design of experiments in ornithology, pp. 1–63. In: Johnston, R.F. (ed.). *Current Ornithology Vol. 2*. Plenum Press, New York.

James, F.C. and Shugart, H.H. (1970). A quantitative method of habitat description. *Audubon Field Notes* **24**, 727–736.

James, P.C. and Robertson, H.A. (1985). The use of playback recordings to detect and census nocturnal burrowing seabirds. *Seabird* **7**, 18–20.

Järvinen, O. and Väisänen, R.A. (1983a). Correction coefficients for line transect censuses of breeding birds. *Ornis Fennica* **60**, 97–104.

Järvinen, O. and Väisänen, R.A. (1983b). Confidence limits for estimates of population density in line transects. *Ornis Scandinavica* **14**, 129–134.

Jenkins, R.K.B., Buckton, S.T. and Ormerod, S.J. (1995). Local movements and population density of Water Rails *Rallus aquaticus* in a small inland reedbed. *Bird Study* **42**, 82–87.

JNCC. (1996). *Birds of Conservation Importance*. Press release. Peterborough, 31 May 1996.

Johnsgaard, P.A. (1994). *Arena birds: Sexual Selection and Behaviour*. Smithsonian Institution Press, Washington.

Johnson, D.H. (1980). The comparison of usage and availability measurements for evaluating resource preference. *Ecology* **61**, 65–71.

Jolly, G.M. (1965). Explicit estimates from capture–recapture data with both death and immigration-stochastic model. *Biometrika* **52**, 225–247.

Jørgensen, A.F. and Nøhr, H. (1996). The use of satellite images for mapping of landscape and biological diversity in the Sahel. *International Journal of Remote Sensing* **17**, 91–109.

Jouventin, P. and Weimerskirch, H. (1990). Satellite tracking of Wandering Albatrosses. *Nature* **343**, 746–748.

Kaiser, von Andreas and Bauer, H-G. (1994). Field experiments to determine the size of breeding populations by means of capture-recapture and visual/acoustical census methods. *Vogelwarte* **37**, 206–231.

Kanyamibwa, S., Schierer, A., Pradel, R. and Lebreton, J.D. (1990). Changes in adult annual survival rates in a western European population of the White Stork *Ciconia ciconia*. *Ibis* **132**, 27–35.

Kendeigh, S.C. (1944). Measurement of bird populations. *Ecological Monographs* **14**, 67–106.

Kenward, R.E. (1987). *Wildlife Radio-tagging: Equipment, Field Techniques and Data Analysis*. Academic Press, London.

Kenward, R.E., Robertson, P.A. and Coates, A.S. (1993). Techniques for radio-tagging pheasant chicks. *Bird Study* **40**, 51–54.

Kerlinger, P. (1989). *Flight Strategies of Migrating Hawks*. University of Chicago Press, Chicago.

Kirby, J.S. (1987). *Birds of Estuaries Enquiry – Instructions to Counters*. BTO, Tring, Herts.

Kirby, J.S. (1990). *A Guide to Birds of Estuaries Enquiry Counting Procedure During the 1982/83 to 1988/89 Period, and Recommendations for the Future*. BTO, Tring, Herts.

Koen, J.H. (1988). A census technique for Afromontane Forest Bird Communities. *Suid-Afrikaanse Bosboutydskrif* **145**, 39–41.

Komdeur, J., Bertelsen, J. and Cracknell, G. (1992). *Manual for Aeroplane and Ship Surveys of Waterfowl and Seabirds*. IWRB Special Publications No. 19. Slimbridge, UK.

Koskimies, P. and Väisänen, R.A. (1991). *Monitoring Bird Populations: a Manual of Methods Applied in Finland*. Finnish Museum of Natural History, Helsinki, Finland.

Lack, P. (1986). *The Atlas of Wintering Birds in Britain and Ireland*. T & AD Poyser, Calton.

Lambin, E.F. and Ehrlich, D. (1996). The surface temperature-vegetation index space for land cover and land-cover change analysis. *International Journal of Remote Sensing* **17**, 463–487.

Lavers, C.P. and Haines-Young, R.H. (1996). The pattern of Dunlin *Calidris alpina* distribution and adundance in relation to habitat variation in the Flow Country of northern Scotland. *Bird Study* **43**, 231–239.

Leslie, P.H. (1945). On the use of matrices in certain population mathematics. *Biometrika* **33**, 183–212.

Liechti, F., Bruderer, B. and Paproth, H. (1995). Quantification of nocturnal bird migration by moonwatching: comparison with radar and infrared observations. *Journal of Field Ornithology* **66**, 457–652.

Lincoln, F.C. (1930). Calculating waterfowl abundance on the basis of banding returns. *USDA Circular* **118**, 1–4.

Lindén, H., Helle, E., Helle, P. and Wikman, M. (1996). Wildlife triangle scheme in Finland: methods and aims for monitoring wildlife populations. *Finnish Game Research* **49**, 4–11.

Lindén, H. and Rajala, P. (1981). Fluctuations in long-term trends in the relative densities of tetraonid populations in Finland, 1964–1977. *Finnish Game Research* **39**, 13–24.

Lloyd, C., Tasker, M.L. and Partridge, K. (1991). *The Status of Seabirds in Britain and Ireland*. T & AD Poyser, Calton.

Lovvorn, J.R. (1989). Distributional responses of Canvasback Ducks to weather and habitat change. *Journal of Applied Ecology* **26**, 113–130.

Lowery, G.H. and Newman, R.J. (1966). A continent wide view of bird migration on four nights in October. *Auk* **83**, 547–586.

MacArthur, R.H. and MacArthur, J.W. (1961). On bird species diversity. *Ecology* **42**, 594–598.

MacArthur, R.H., MacArthur, J.W. and Preer, J. (1962). On bird species diversity: II. Prediction of bird census from habitat measurements. *American Naturalist* **96**, 167–174.

Maddock, M.N. and Geering, D.J. (1994). Effect of patagial tags on cattle egrets. *Corella* **18**, 1–7.

Marchant, J.H. (1983). *BTO Common Birds Census Instructions*. BTO, Tring, Herts.

Marchant, J.H., Hudson, R., Carter, S.P. and Whittington, P. (1990). *Population Trends in British Breeding Birds*. BTO, Tring, Herts.

Margules, C.R. and Austin, M.P. (1994). Biological models for monitoring species decline: the construction and use of data-bases. *Philosophical Transactions of the Royal Society, London* **B 344**, 69–75.

Margules, C.R. and Redhead, T.D. (1996). *Guidelines for Using the BioRap Methodology and Tools*. CSIRO for the World Bank, Australia.

Marion, W. R., O'Meara, T.E. and Maehr, D.S. (1981). Use of playback recordings in sampling elusive or secretive birds. *Studies in Avian Biology* **6**, 81–85.

Marquiss, M. (1989). Grey Herons *Ardea cinerea* breeding in Scotland: numbers' distribution, and census techniques. *Bird Study* **36**, 181–191.

Marquiss, M., Newton, I. and Ratcliffe, D.A. (1978). The decline of the Raven, *Corvus corax*, in relation to afforestation in southern Scotland and northern England. *Journal of Applied Ecology* **15**, 129–144.

Marsden, S.J. (1999). Estimation of parrot and hornbill densities using a point count distance sampling method. *Ibis* **141**, 377–390.

Marsden, S.J., Jones, M.J., Linsley, M.D., Mead, C. and Hounsome, M.V. (1997). The conservation status of the restricted range lowland birds of Indonesia. *Bird Conservation International* **7**, 213–233.

Martella, M.B. and Navarro, J.L. (1992). Capturing and marking Greater Rheas. *Journal of Field Ornithology* **63**, 117–120.

Massa, R. and Fedrigo, A. (1989). A new approach for compiling a winter bird atlas by means of point counts. *Annales Zoologici Fennici* **26**, 207–212.

Matthysen, E. (1989). Nuthatch *Sitta europaea* demography, beech mast, territoriality. *Ornis Scandinavica* **20**, 278–282.

McCracken, D.I. (1994). Extensive farming systems in Europe: an initial assessment of the situation between the United Kingdom and the Republic of Ireland. Report to the Institute for European Environmental Policy, London.

McKinnon, J. and Phillips, K. (1993). *A Field Guide to the Birds of Borneo, Sumatra, Java and Bali*. Oxford University Press, Oxford.

Mead, C. (1987). *Owls*. Whittet Books, London.

Meek, E.R., Booth, C.J., Reynolds, P. and Ribbands, B. (1983). Breeding skuas in Orkney. *Seabird* 7, 21–29.

Meltofte, H., Blew, J., Frikke, J., Rösner, H.-U. and Smit, C.J. (1994). *Numbers and Distribution of Waterbirds in the Wadden Sea. Results and evaluation of 36 Simultaneous Counts in the Dutch-German-Danish Wadden Sea 1980–1991.* IWRB Publications 34/Wader Study Group Bulletin 74. Special Issue, pp. 192.

Meyburg, B.U. and Chancellor, R.D. (eds.) (1989). *Raptors in the modern world.* World Working Group on Birds of Prey and Owls: Berlin.

Meyburg, B.U. and Chancellor, R.D. (eds.) (1994). *Raptor Conservation Today.* Pica Press, for the World Working Group on Birds of Prey and Owls: Berlin.

Miller, R.G. (1974). The Jackknife – a review. *Biometrika* 61, 1–15.

Miller, K.V. and Conroy, M.J. (1990). Spot satellite imagery for mapping Kirkland's Warbler wintering habitat in the Bahamas. *Wildlife Society Bulletin* 18, 252–257.

Millsap, B.A. and LeFranc, M.N., Jr. (1988). Road transects for raptors: how reliable are they? *Journal of Raptor Research* 22, 8–16.

Mosher, J.A., Fuller, M.R. and Kopeny, M. (1990). Surveying woodland raptors by broadcast of conspecific vocalisations. *Journal of Field Ornithology* 61, 453–461.

Moss, R. and Oswald, J. (1985). Population dynamics of Capercaillie in a north-east Scottish glen. *Ornis Scandinavica* 16, 229–238.

Moyer, D. (1993). A preliminary trial of territory mapping for estimating bird densities in afromontane forests. *Proceedings of the 8th Pan-African Ornithological Congress* 302–311.

Mudge, G.P. (1988). An evaluation of current methodology for monitoring changes in the breeding populations of Guillemots *Uria aalge. Bird Study* 35, 1–9.

Munn, C.A. and Terborgh, J.W. (1979). Multi-species territoriality in Neotropical foraging flocks. *Condor* 81, 338–347.

Murray, S. and Wanless, S. (1986). The status of the Gannet in Scotland 1984–85. *Scottish Birds* 14, 74–85.

NCC (1990). *Handbook for Phase 1 Habitat Survey: a Technique for Environmental Audit.* England Field Unit, Nature Conservancy Council, Peterborough.

Nettleship, D.N. (1976). Census techniques for seabirds of Arctic and Eastern Canada. *Canadian Wildlife Service Occasional Papers* 25, 1–33.

Neu, C.W., Byers, C.R. and Peek, J.M. (1974). A technique for analysis of utilisation availability data. *Journal of Wildlife Management* 38, 541–545.

Newton, I. (1986). *The Sparrowhawk.* T & AD Poyser, Calton.

Newton, I. (1988). A key factor analysis of a Sparrowhawk population. *Oecologia* 76, 588–596.

Newton, I. and Chancellor, R.D. (eds.) (1985). *Conservation Studies on Raptors.* ICBP Technical Publication No. 5, Cambridge.

Newton, I., Wyllie, I. and Mearns, R. (1986). Spacing of Sparrowhawks in relation to food supply. *Journal of Animal Ecology* 55, 361–370.

Nichols, J.D., Noon, B.R., Stokes, S.L. and Hines, J.E. (1981). Remarks on the use of mark–recapture methodology in estimating avian population size. *Studies in Avian Biology* 6, 121–136.

Nisbet, I.C.T. (1959). Calculation of flight direction of birds observed crossing the face of the moon. *Wilson Bulletin* 71, 237–243.

Nøhr, H. and Jørgensen, A.F. (1997). Mapping of biological diversity in Sahel by means of satellite image analyses and ornithological surveys. *Biodiversity and Conservation* 6(4), 545–566.

O'Connor, R.J. and Mead, C.J. (1984). The Stock Dove in Britain, 1930–80. *British Birds* 77, 181–201.

O'Connor, R.J. and Shrubb, M. (1986). *Farming and Birds.* Cambridge University Press, Cambridge.

Ogilvie, M.A. (1986). The Mute Swan *Cygnus olor* in Britain 1983. *Bird Study* 33, 121–137.

Ormerod, S.J., Tyler, S.J., Pester, SJ. and Cross, A.V. (1988). Censusing distribution and population of birds along upland rivers using measured ringing effort: a preliminary study. *Ringing and Migration* 9, 71–82.

Osborne, P.E. and Tigar, B. (1992). Interpreting bird atlas data using logistic models: an example from Lesotho, South Africa. *Journal of Applied Ecology* 29, 55–62.

Österlöf, S. and Stolt, B.-O. (1982). Population trends indicated by birds ringed in Sweden. *Ornis Scandinavica* **13**, 135–140.

Otis, D.L., Burnham, K.P., White, G.C. and Anderson, D.R. (1978). Statistical inference from capture data on closed animal populations. *Wildlife Monographs* **62**, 1–135.

Owen, M. (1971). The selection of feeding sites by White-fronted Geese in winter. *Journal of Applied Ecology* **8**, 905–917.

Owen, M., Atkinson-Willes, G.L. and Salmon, D.G. (1980). *Wildfowl in Great Britain*. Cambridge University Press, Cambridge.

Owen, M. and Black, J.M. (1989). Factors affecting the survival of Barnacle Geese on migration from the breeding grounds. *Journal of Animal Ecology* **58**, 603–617.

Pain, D.J., Hill, D. and McCracken, D.I. (1997). Impact of agricultural intensification of pastoral systems on bird distribution in Britain, 1970–1990. *Agriculture, Ecosystems and Environment* **64**, 19–32.

Palmer, R.S. (ed.) (1962 onwards). *Handbook of North American Birds*. Vols. I–V. Yale University Press, New Haven and London.

Parker, T.A. III (1991). On the use of tape-recorders in avifaunal surveys. *Auk* **108**, 443–444.

Parrinder, E.D. (1989). Little Ringed Plovers *Charadrius dubius* in Britain in 1984. *Bird Study* **36**, 147–153.

Peach, W. and Baillie, S.R. (1989). Population changes on constant effort sites, 1987–1988. *BTO News* **161**, 12–13.

Peach, W.J., Baillie, S.R. and Balmer, D.E. (1998). Long term changes in the abundance of passerines in Britain and Ireland as measured by constant effort mist-netting. *Bird Study* **45**, 257–275.

Peach, W.J., Buckland, S.T. and Baillie, S.R. (1990). Estimating survival rates using mark-recapture data from multiple ringing sites. *Ring* **13**, 87–102.

Petty, S.J. and Avery, M.I. (1990). Forest Bird Communities. *Forestry Commission, Occasional Paper* **26**, 1–110.

Pienkowski, M.W., Stroud, D.A. and Bignal, E.M. (1990). Estimating bird numbers and distributions in extensive survey areas using remote sensing. In: Stastny, K. and Bejcek, V. (eds.). *Bird Census and Atlas Studies*. Proceedings of XI International Conference on Bird Census and Atlas Work. Prague 1990.

Piersma, T. and Ntiamoa-Baidu, Y. (1995). *Waterbird Ecology and the Management of Coastal Wetlands in Ghana*. Netherlands Institute for Sea Research, Texel.

Pihl, S. and Frikke, J. (1992). Counting birds from aeroplane. In: Komdeur, J., Bertelsen, J. and Cracknell, G. (eds.). *Manual for Aeroplane and Ship Survey of Waterfowl and Seabirds*. IWRB Special Publication 19. Slimbridge. UK.

Pithon, J.A. and Dytham, C. (1999). Census of the British Ring-necked Parakeet *Psittacula krameri* population by simultaneous counts of roosts. *Bird Study* **46**, 112–115.

Pollock, K.H. (1981). Capture–recapture models: a review of current methods, assumptions and experimental design. *Studies in Avian Biology* **6**, 426–435.

Pomeroy, D. (1989). Using East African bird atlas data for ecological studies. *Annales Zoologici Fennici* **26**, 309–314.

Pomeroy, D.E. (1992). *Counting Birds*. African Wildlife Foundation, Nairobi.

Pomeroy, D. and Dranzoa, C. (1997). Methods of studying the distribution, diversity and abundance of birds in East Africa – some quantitative approaches. *African Journal of Ecology* **35**, 110–123.

Porter, J.M. and Coulson, J.C. (1987) Long-term changes in recruitment to the breeding group, and the quality of recruits at a Kittiwake *Rissa-Tridactyla* colony. *Journal of Animal Ecology* **56**, 675–689.

Porter, R.F. and Beaman, M.A.S. (1985). *A Resumé of Raptor Migration in Europe and the Middle East. Conservation Studies on Raptors*. Ed. Newton, I. and Chancellor, R.D. pp. 237–242. ICBP Technical Publication No. 5, Cambridge.

Potts, G.R. (1986). *The Partridge: Pesticides, Predation and Conservation*. Collins, London.

Poulsen, B.O. (1994). Mist-netting as a census method for determining species richness and abundances in an Andean Cloud forest bird community. *Le Gerfaut* **84**, 39–49.

Poulsen, B.O., Krabbe, N., Frølander, A., Hinojosa, M.B. and Quiroga, C.O. (1997). A rapid assessment of Bolivian and Ecuadorian montane avifaunas using 20-species lists: efficiency, biases and data gathered. *Bird Conservation International* **7**, 53–67.

Pöysä, H. (1984). Temporal and spatial dynamics of waterfowl populations in a wetland area – a community ecological approach. *Ornis Fennica* **61**, 99–108.

Pöysä, H. (1996). Population estimates and the timing of waterfowl censuses. *Ornis Fennica* **73**, 60–68.

Prater, A.J. (1979). Trends in accuracy of counting birds. *Bird Study* **26**, 198–200.

Prater, A.J. (1981). *Estuary Birds of Britain and Ireland*. T & AD Poyser, Calton.

Prater, A.J. (1989). Ringed plover *Charadrius hiaticula* breeding population of the United Kingdom in 1984. *Bird Study* **36**, 154–161.

Price, J., Droege, S. and Price, A. (1995). *The Summer Atlas of North American Birds*. Academic Press, New York.

Pyle, P., Nur, N. and DeSante, D.F. (1994). Trends in nocturnal migrant landbird populations at southeast Farallon Island, California, 1968–1992. *Studies in Avian Biology* **15**, 58–74.

Quade, D. (1979). Using weighted rankings in the analysis of complete blocks with additive block effects. *Journal of American Statistics Association* **74**, 680–683.

Ralph, C.J. and Scott, J.M. (eds.) (1981). *Estimating the Number of Terrestrial Birds: Studies in Avian Biology no. 6*. Cooper Ornithological Society, Lawrence, Kansas, USA.

Ralph, C.J., Geupel, G.R., Pyle, R., Thomas, E. and DeSante, D.F. (1993). *Handbook of Field Methods for Monitoring Landbirds*. Gen. Tech. Rep. PSW-GTR-144. Pacific Southwest Research Station, Forest Service, U.S. Department of Agriculture, Albany, CA.

Rapold, C., Kersten, M. and Smit, C. (1985). Errors in large scale shorebird counts. *Ardea* **73**, 13–24.

Ratcliffe, N., Vaughan, D., Whyte, C. and Shepherd, M. (1998). Development of playback census methods for Storm Petrels *Hydrobates pelagicus*. *Bird Study* **45**, 302–312.

Rebecca, G.W. and Bainbridge, I.P. (1998). The breeding status of the Merlin *Falco columbarius* in Britain in 1993–94. *Bird Study* **45**, 172–187.

Redpath, S.M. (1994). Censusing Tawny Owls *Strix aluco* by the use of imitation calls. *Bird Study* **41**, 192–198.

Reed, T.M. and Fuller, R.J. (1983). Methods used to assess populations of breeding waders on machair in the Outer Hebrides. *Wader Study Group Bulletin* **39**, 14–16.

Reed, T.M., Barrett, J.C., Barrett, C. and Langslow, D.R. (1984). Diurnal variability in the detection of dunlin *Calidris alpina*. *Bird Study* **31**, 245–246.

Reed, T.M., Barrett, C., Barrett, J., Hayhow, S. and Minshull, B. (1985). Diurnal variability in the detection of waders on their breeding grounds. *Bird Study* **32**, 71–74.

Remsen, J.V. Jr. (1994). Use and misuse of bird lists in community ecology and conservation. *Auk* **111**, 225–227.

Remsen, J.V. and Good, D.A. (1996). Misuse of data from mist-net captures to assess relative abundance in bird populations. *Auk* **113**, 381–398.

Reynolds, C.M. (1979). The heronries census: 1972–1977 population changes and a review. *Bird Study* **26**, 7–12.

Reynolds, R.T., Scott, J.M. and Nussbaum, R.A. (1980). A variable circular plot method for estimating bird numbers. *Condor* **82**, 309–313.

Richardson, M.G. (1990). The distribution and status of Whimbrel *Numenius p. phaeopus* in Shetland and Britain. *Bird Study* **37**, 61–88.

Robbins, C.S. (1981). Effect of time of day on bird activity. *Studies in Avian Biology* **6**, 275–286.

Robbins, C.S. (1990). Use of breeding bird atlases to monitor population change. pp. 18–22. In: Sauer, J.R. and Droege, S. (eds.). (1990). *Survey Designs and Statistical Methods for the Estimation of Avian Population Trends*. U.S. Fish and Wildlife Service, Biological Report **90** (1).

Robbins, C.S., Bystrak, D. and Geissler, P.H. (1986). The Breeding Bird Survey: its first fifteen years, 1965–1979. *United States Department of the Interior, Fish and Wildlife Service, resource publication* **157**, 1–196.

Robbins, C.S., Droege, S. and Sayer, J.R. (1989). Monitoring bird populations with Breeding Bird Survey and atlas data. *Annales Zoologici Fennici* **26**, 279–304.

Robertson, A., Simmons, R.E., Jarvis, A.M. and Brown, C.J. (1995). Can bird atlas data be used to estimate population size? A case study using Namibian endemics. *Biological Conservation* **71**, 87–95.

Robertson, P.A., Woodburn, M.I.A., Bealey, C.E., Ludolf, I.C. and Hill, D.A. (1990). *Pheasants and Woodlands: Habitat Selection, Management and Conservation.* Report to the Forestry Commission. The Game Conservancy, Fordingbridge.

Rodwell, J.S. (ed.) (1991). *British Plant Communities: vol. 1. Woodlands and Scrub.* Cambridge University Press, Cambridge.

Rolstad, J. and Wegge, P. (1987). Distribution and size of Capercaillie leks in relation to old forest fragmentation. *Oecologia* **72**, 389–394.

Root, T. (1988). *Atlas of Wintering North American Birds. An Analysis of Christmas Bird Count Data.* University of Chicago Press, Chicago.

Rose, P. (1990). *Manual for International Waterfowl Census Coordinators.* International Waterfowl and Wetlands Research Bureau, Slimbridge.

Rose, P.M. and Scott, D.A. (1994). *Waterfowl population estimates.* IWRB Publication 29. International Waterfowl and Wetlands Research Bureau, Slimbridge.

Rose, P.M. and Taylor, V. (1993). *Western Palaearctic and S. West Asia Waterfowl Census, 1993.* International Waterfowl Research Bureau, Slimbridge.

Roseberry, J.L. and Hao, Q. (1996). Interactive computer program for landscape-level habitat analysis. *Wildlife Society Bulletin* **24**, 340–341.

Rotenberry, J. (1985). The role of habitat in avian community composition: physiognomy or floristics? *Oecologia* **67**, 213–217.

Rotenberry, J.T. and Wiens, J.A. (1980). Habitat structure, patchiness, and avian communities in North American steppe vegetation: a multivariate analysis. *Ecology* **61**, 1228–1250.

Rotenberry, J.T. and Wiens, J.A. (1985). Statistical power analysis and community wide patterns. *The American Naturalist* **125**, 164–168.

Rumble, M.A. and Flake, L.D. (1982). A comparison of two waterfowl brood survey techniques. *Journal of Wildlife Management* **46**, 1048–1053.

Russell, K.R., Mizrahi, D.S. and Gauthreaux, S.A. Jr. (1998). Large-scale mapping of purple martin pre-migratory roosts using WSR-88D weather surveillance radar. *Journal of Field Ornithology* **69**, 316–325.

Ryan, P.G. and Cooper, J. (1989). The distribution and abundance of aerial seabirds in relation to Antarctic krill in the Pryds Bay region, Antarctica, during late Summer. *Polar Biology* **10**, 199–209.

Sader, S.A., Powell, G.V.N. and Rappole, J.H. (1991). Migratory bird habitat monitoring through remote sensing. *International Journal of Remote Sensing* **12**, 363–372.

Sage, B.L. and Vernon, J.D.R. (1978). The 1975 National survey of Rookeries. *Bird Study* **25**, 64–86.

Salmon, D.G. (1989). In: Prys-Jones, R. and Kirby, J. (eds.). *Wildfowl and Wader Counts 1988–89.* Wildfowl and Wetlands Trust, Slimbridge.

Sample, V.A. (1994). *Remote Sensing and GIS in Ecosystem Management.* Island Press, Washington, D.C.

Sauer, J.R. and Droege, S. (1990). *Survey Designs and Statistical Methods for the Estimation of Avian Population Trends.* US Fish and Wildlife Service. Biological Report 90(1).

Saunders, D.A. and Wooler, R.D. (1988). Consistent individuality of voice in birds as a management tool. *Emu* **88**, 25–32.

Schuster, S. (1975). Schätzfehler bei Wasservogel-"Zählungen". *Ornithologische Mitteilungen* **27**, 250.

Schwaller, M.R., Olser, C.E., Zhenqui Ma, Zhiliang Zhu and Dahmer, P. (1989). A remote sensing analysis of Adelie penguin rookeries. *Remote Sensing of Environment* **28**, 199–206.

Scott, M.J. and Ramsay, F.L. (1981). Length of count period as possible source of bias in estimating bird densities. *Studies in Avian Biology* **6**, 409–413.

Scott, M.J., Ramsay, F.L. and Kepler, C.B. (1981). Distance estimation as a variable in estimating bird numbers. *Studies in Avian Biology* **6**, 334–341.

Scott, P. (1981). *Variation of Bill-Markings of Migrant Swans Wintering in Britain.* The Wildfowl Trust, Slimbridge.

Seabird Group/NCC (1988). *Seabird Colony Register: Recommended Methods for Counting Breeding Seabirds.* Seabird Group/NCC, Peterborough, U.K.

Seber, G.A.F. (1965). A note on the multiple-recapture census. *Biometrika* **52**, 249.

Seber, G.A.F. (1973). *The Estimation of Animal Abundance*. Hafner, New York and Griffin, London.

Sharrock, J.T.R. (1976). *The Atlas of Breeding Birds in Britain and Ireland*. T & AD Poyser, Calton.

Shawyer, C.R. (1987). *The Barn Owl in Britain: Its Past, Present and Future*. The Hawk Trust, London.

Shrubb, M. and Lack, P.C. (1991). The numbers and distribution of Lapwings *V. vanellus* nesting in England and Wales in 1987. *Bird Study* **38**, 20–38.

Sitters, H. (ed.). (1988). *Tetrad Atlas of the Breeding Birds of Devon*. Devon Birdwatching and Preservation Society, Yelverton.

Skov, H., Durinck, J., Leopold, M.F. and Tasker, M.L. (1995). *Important Bird Areas for Seabirds in the North Sea*. BirdLife International, Cambridge.

Smith, J.M. (1988). Landsat TM study of afforestation in northern Scotland and its impact on breeding bird populations, pp. 1369–1370. In: Gvyenne, T.D. and Hunt, J.J. (eds.). *Remote sensing. Proceedings IGARSS '88 symposium, Edinburgh. Vol. 3*. European Space Agency, ESTEC, Noordwijk, ESA, SP–284.

Smith, K.W. (1983). The status and distribution of waders breeding on lowland wet grasslands in England and Wales. *Bird Study* **30**, 177–192.

Smith, N.G. (1980). Hawk and vulture migration in the Neotropics. pp. 51–65. In: Keast, A. and Morton, E.S. (eds.). *Migrant birds in the Neotropics: Ecology, Behaviour and Conservation*. Smithsonian Institution Press, Washington D.C.

Smith, N.G. (1985). Dynamics of the transishmian migration of raptors between Central and South America. In: Newton, I. and Chancellor, R.D. (eds) *Conservation Studies on Raptors*. pp. 237–242. ICBP Technical Publication No. 5, Cambridge.

Southern, H.N. and Lowe, V.P.W. (1968). Pattern of distribution of prey and predation in Tawny Owls. *Journal of Animal Ecology* **37**, 75–97.

Southwood, T.R.E. (1978). *Ecological Methods*. 2nd Edn. Chapman and Hall, London.

SOVON (1987). *Atlas van de Nederlandse Vogels*. SOVON, Arnhem.

Spellerberg, I. (1991). *Monitoring Ecological Change*. Cambridge University Press, Cambridge.

Stattersfield, A.J., Crosby, M.C., Long, A.J. and Wege, D.C. (1998). *Endemic Bird Areas of the World: Priorities for Biodiversity Conservation*. BirdLife International, Cambridge.

Stoms, D.M. and Estes, J.E. (1993). A remote sensing research agenda for mapping and monitoring biodiversity. *International Journal for Remote Sensing* **14**, 1839–1860.

Stone, B.H., Sears, J., Cranswick, P.A., Gregory, R.D., Gibbons, D.W., Rehfisch, M.M., Aebischer, N.J. and Reid, J.B. (1997). Population estimates of birds in the United Kingdom. *British Birds* **90**, 1–22.

Stone, C.J., Webb, A., Barton, C., Ratcliffe, N., Camphuysen, C.J., Reed, T.C., Tasker, M.L. and Pienkowski, M.W. (1995). *An Atlas of Seabird Distribution in North-west European Waters*. JNCC, Peterborough.

Stowe, T.J. (1982). Recent population trends in cliff-breeding seabirds in Britain and Ireland. *Ibis* **124**, 502–510.

Stowe, T.J. and Hudson, A.V. (1988). Corncrake studies in the western isles. *RSPB Conservation Review* **2**, 38–42.

Sutherland, W.J. (ed.) (1996). *Ecological Census Techniques: a Handbook*. Cambridge: Cambridge University Press.

Svensson, S. (1993). Bird monitoring with breeding bird censuses and migration counts [Swedish with English summary]. *Vår Fuglefauna Suppl.* **1**, 3–12.

Swenson, J.E. (1991). Evaluation of a density index for territorial male Hazel Grouse *Bonasa bonasia* in spring and autumn. *Ornis Fennica* **68**, 57–65.

Tappe, P.A., Whiting, R.M. and George, R.R. (1989). Singing-ground surveys for Woodcock in East Texas. *Wildlife Society Bulletin* **17**, 36–40.

Tapper, S. (1989). The 1989/90 shooting season. pp. 28–34. In: Nodder, C. (ed.) *The Game Conservancy Review of 1989*. Game Conservancy, Fordingbridge.

Tasker, M.L., Hope-Jones, P., Dixon, T. and Blake, B.F. (1984). Counting seabirds at sea from ships: a review of methods employed and a suggestion for a standardized approach. *Auk* **101**, 567–577.

Taylor, K. (1982). *BTO Waterways Bird Survey Instructions*. British Trust for Ornithology, Thetford.

Taylor, K., Hudson, R. and Horne, G. (1988). Buzzard breeding distribution and abundance in Britain and Ireland in 1983. *Bird Study* **35**, 109–118.

Terborgh, J., Robinson, S.K., Parker, T.A. III, Munn, C.A. and Pierpont, N. (1990). Structure and organisation of an Amazonian forest bird community. *Ecological Monographs* **60**, 213–238.

Thomas, L., Laake, J.L., Derry, J.F., Buckland, S.T., Borchers, D.L., Anderson, D.R., Burnham, K.P., Strindberg, S., Hedley, S.L., Burt, M.L., Marques, F., Pollard, J.H. and Fewster, R.M. (1998). *Distance 3.5*. Research Unit for Wildlife Population Assessment, University of St Andrews, UK.

Thompson, J.J. (1989). A comparison of some avian census techniques in a population of Lovebirds at Lake Naivasha, Kenya. *African Journal of Ecology* **27**, 157–166.

Thompson, K.R. and Rothery, P. (1991). A census of Black-browed Albatross *Diomedea melanophyrs* population on Steeple Jason Island, Falkland Islands. *Biological Conservation* **56**, 39–48.

Tucker, G.M. and Heath, M.F. (1994). *Birds in Europe: their Conservation Status*. BirdLife International, Cambridge.

Udvardy, M.D.F. (1981). An overview of grid-based atlas works in ornithology. *Studies in Avian Biology* **6**, 103–109.

Underhill, L.G. and Fraser, M.W. (1989). Bayesian estimate of the number of Malachite Sunbirds feeding at an isolated and transient nectar resource. *Journal of Field Ornithology* **60**, 382–387.

Underhill, M.C., Gittings, T., Callaghan, D.A., Hughes, B., Kirby, J. and Delany, S. (1998). Status and distribution of breeding Common Scoters *Melanitta nigra nigra* in Britain and Ireland in 1995. *Bird Study* **45**, 146–156.

Verner, J. (1985). Assessment of counting techniques, pp. 247–301. In: Johnston, R.F. (ed.). *Current Ornithology Vol. 2*. Plenum Press, New York.

Verner, J. and Milne, K.A. (1989). Coping with sources of variability when monitoring population trends. *Annales Zoologici Fennici* **26**, 191–199.

Vinicombe, K. (1982). Breeding and population of the Little Grebe. *British Birds* **75**, 204–218.

Viñuela, J. (1997). Road transects as a large-scale census method for raptors: the case of the Red Kite *Milvus milvus* in Spain. *Bird Study* **44**, 155–165.

Walsh, P.M., Halley, D.J., Harris, M.P., del Nevo, A., Sim, I.M.W. and Tasker, M.L. (1995). *Seabird Monitoring Handbook for Britain and Ireland*. JNCC/RSPB/ITE/Seabird Group.

Wanless, S. and Harris, M.P. (1984). Effects of date on counts of nests of Herring and Lesser Black-backed Gulls. *Ornis Scandinavica* **15**, 89–94.

Waters, R., Cranswick, P., Smith, K.W. and Stroud, D. (1996). Wetland Bird Survey. In: Cadbury, C.J. (ed.) *1996 RSPB Conservation Review*.

Watson, A., Payne, S. and Rae, R. (1989). Golden Eagles *Aquila chrysaetos:* land use and food in north-east Scotland. *Ibis* **131**, 336–348.

Webb, A. and Durinck, J. (1992). Counting birds from ship. In: Komdeur, J., Bertelsen, J. and Cracknell, G. (eds.) *Manual for Aeroplane and Ship Surveys of Waterfowl and Seabirds*, IWRB Special Publication No. 19 IWRB/JNCC/Ornis Consult A/S Ministry of the Environment, Copenhagen, Denmark.

Weiss, N.T., Samuel, M.D., Rusch, D.H. and Caswell, F.D. (1991). Effects of resighting errors on capture-resight estimates for neck-banded Canada Geese. *Journal of Field Ornithology* **62**, 464–473.

Welsh, D.A. (1989). A report on breeding bird atlases in Canada. *Annales Zoologici Fennici* **26**, 305–308.

Whilde, A. (1985). The 1984 all-Ireland tern survey. *Irish Birds* **3**, 1–32.

White, F. (1983). *The Vegetation of Africa*. UNESCO, Paris.

Whitman, A.A., Hagan, J.M.H (III) and Nicholas, V.L.B (1997). A comparison of two bird survey techniques used in a subtropical forest. *The Condor* **99**, 965–995.

Whittaker, R.H. (1977). Evolution of species diversity in land communities. *Evolutionary Biology* **10**, 1–67.

Wickstrom, D.C. (1982). Factors to consider in recording avian sounds. In: Kroodsma, D.E. and Miller, E.H. (eds.) *Acoustic Communication in Birds*. Academic Press, New York.

Wiens, J.A. (1969). An approach to the study of ecological relationships among grassland birds. *Ornithological Monographs* **8**, 1–93.

Wiens, J.A. (1973). Pattern and process in grassland bird communities. *Ecological Monographs* **43**, 237–270.

Wiens, J.A. (1981). Scale problems in avian censusing. *Studies in Avian Biology* **6**, 513–521.

Wiens, J.A. (1985). Habitat selection in variable environments: shrubsteppe birds, pp. 227–251. In: Cody, M.L. (ed.). *Habitat Selection in Birds*. Academic Press, New York.

Wiens, J.A. (1989). *The Ecology of Bird Communities, vols. 1 and 2*. Cambridge University Press, Cambridge.

Wiens, J.A. and Rotenberry, J.T. (1985). Response of breeding passerine birds to rangeland alteration in a North American shrubsteppe locality. *Journal of Applied Ecology* **22**, 655–668.

Wiley, R.H. (1980). Multispecies antbird societies in lowland forests of Surinam and Ecuador: stable membership and foraging differences. *Journal of Zoology London* **191**, 127–145.

William, A.B. (1936). The composition and dynamics of a beech–maple climax community. *Ecological Monographs* **6**, 317–408.

Williams, P.H., Gibbon, D., Margules, C., Rebalo, A., Humphries, C. and Pressey, R. (1996). A comparison of richness hotspots, rarity hotspots and complementarity areas for conserving the diversity of British birds. *Conservation Biology* **10**, 155–174.

Williamson, K. (1964). Bird census work in woodland. *Bird Study* **11**, 1–22.

Williamson, K. (1968). Buntings on a barley farm. The bird community of farmland. *Bird Study* **15**, 34–37.

Wilson, H.J. (1982). Movements, home ranges and habitat use of wintering Woodcock in Ireland, pp. 168–178. In: Dwyer, T.J. and Storm, G.L. (eds.). *Papers of the Seventh Woodcock Symposium. Wildlife Research Report no. 14*. United States Department of the Interior: Fish and Wildlife Service, Washington D.C.

Woolhead, J. (1987). A method for estimating the number of breeding pairs of Great Crested Grebes *Podiceps cristatus* on lakes. *Bird Study* **34**, 82–86.

Wooller, R.D., Bradley, J.S., Skira, I.J. and Serventy, D.L. (1990) Reproductive success of Short-tailed Shearwaters *Puffinus tenuirostris* in relation to their age and breeding experience. *Journal of Animal Ecology* **59**, 161–170.

Wormell, P. (1976). The Manx Shearwaters of Rhum. *Scottish Birds* **9**, 103–118.

Yalden, D.W. and Yalden, P.E. (1989). The sensitivity of breeding Golden Plovers *Pluvialis apricaria* to human intruders. *Bird Study* **36**, 49–55.

Young, A.D. (1989). Spacing behaviour of visual and tactile-feeding shorebirds in mixed species groups. *Canadian Journal of Zoology* **67**, 2026–2028.

Zar, J. H. (1984). *Biostatistical Analysis*. Prentice-Hall Inc., Englewood Cliffs, N.J.

Species Index

General Index

Note: All references are to **birds** and **Britain**, unless otherwise indicated. Page references for **chapters** are in **bold**